HRW

ALGEBRA ONE
INTERACTIONS
COURSE 1

COOPERATIVE-LEARNING ACTIVITIES

HOLT, RINEHART AND WINSTON
Harcourt Brace & Company

Austin • New York • Orlando • Atlanta • San Francisco • Boston • Dallas • Toronto • London

To the Teacher

HRW Algebra One Interactions Course 1 Cooperative-Learning Activities contain one-page blackline masters for each of the 83 lessons in *HRW Algebra One Interactions Course 1*. These masters provide structured activities for students to perform in small groups or in pairs to reinforce the mathematical content in the lessons. Directions designate specific roles or responsibilities that facilitate student cooperation and participation.

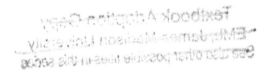

Developmental assistance by B&B Communications West, Inc.

HRW is a registered trademark licensed to Holt, Rinehart and Winston.

Printed in the United States of America

ISBN 0-03-051277-8

3   4   5   6   7   066   00   99   98

TABLE OF CONTENTS

Cooperative-Learning Activity
1.1 Describing Patterns

Group Members: 2

Materials: folders, cubes, buttons, shape blocks, or any small familiar objects

Roles: **Pattern Maker** makes a hidden pattern and describes it so that it can be duplicated

Pattern Matcher attempts to duplicate the pattern without asking questions

Preparation: The goal of this activity is to duplicate a pattern based on a word description without looking at your partner's patterns. Two sample patterns are provided as examples.

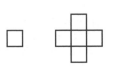

Procedures:

1. The players decide on their roles. Divide the small objects evenly between both partners. Stand the folders up by opening them to about 120° so that neither member can see behind his or her partner's upright folder.

2. The Pattern Maker creates a pattern with the objects hidden from view behind the upright folder.

3. The Pattern Maker describes the pattern so that the Pattern Matcher can duplicate the pattern behind his or her own folder. The Pattern Maker may not look to see what the Pattern Matcher does, and the Pattern Matcher must remain silent, not asking any questions.

4. After the pattern is duplicated, remove the folders and observe both patterns.

5. Switch roles, with the new Pattern Maker creating a different pattern and repeating the above procedures.

6. Discuss which role is harder and explain why.

_____

_____

7. Describe techniques that were especially helpful.

_____

_____

8. Identify some relationships of the patterns that were hard to describe.

_____

_____

Cooperative-Learning Activity
1.2 Pulse Rates

Group Members: 2

Materials: stopwatch or a watch with a second hand and pencil

Responsibilities: Measure and record pulse rates while resting and after doing various activities.

Preparation: The pulse can be defined as the arterial pressure wave generated by the opening and closing of the aortic valve in the heart. As the heart beats faster, the pulse rate increases. The heart rate is equal to the pulse rate. The wave, or pulse rate, can be felt by applying pressure with your fingertip to pulse points found in the neck, inside the elbow, and inside the wrist.

Procedures:

1. Work with a partner to determine each other's pulse rate. Place your index finger and middle finger of one hand on one of the pulse points of your partner. Look at a watch or a clock with a second hand and count the number of beats that you feel in 10 seconds. Record the data in the chart for a resting pulse rate.

2. Switch roles and repeat Procedure 1 to measure your partner's pulse rate.

3. Now measure your partner's pulse rate immediately after he or she completes some physical activity, such as as running in place, fast walking, jogging, or climbing stairs.

4. Switch roles and repeat Procedure 1 after both partners have exercised.

5. Use the equation $h = 6b$, where h is your pulse rate per minute and b is the number of heartbeats for 10 seconds, to find your heart rate.

6. When both partners have completed the activity, review your tables and discuss any similarities and differences.

Member	Activity	Heartbeats per 10 seconds (b)	Heart rate per minute (h)
Example	resting	20	6(20) = 120 beats/min

Cooperative-Learning Activity
1.3 Finding the Least Expensive Situation

Group members: 3

Materials: paper, pencil, calculator, and a ruler

Responsibilities: Create a table and a bar chart from an equation.

Preparation: A gas company is considering billing its customers based on one of the formulas. The monthly charge, c, is equal to the number of cubic feet (ccf) of gas used, f, times a multiplier plus a monthly service charge of $7.00, $8.00, or $9.00.

$$c = 0.323f + 7.00$$
$$c = 0.305f + 8.00$$
$$c = 0.279f + 9.00$$

Procedures: Use the following procedures to determine which customers benefit the most from each situation:

1. Each group member selects a different equation and makes a table to show the monthly bill for 5, 10, 15, 20 ,25, 30, 35, 40, 45, and 50 cubic feet. The table should have two columns: one representing the gas used, f, and one representing the monthly charge, c. Round results to the nearest hundredth.

2. Each member then creates a bar chart from his or her table. To make comparisons between charts easy, group members should decide on uniform scales for the x- and y-axes.

3. As a group, compare the three tables and charts to determine which formula gives the lowest customer cost for an average usage of 30 ccf a month. What is the monthly cost for 30 ccf?

_____

4. Find the formula(s) that give(s) the lowest cost for each value of f listed in your tables.

_____

5. As customers, which formula would your group recommend? What further information do you need in order to make this decision?

_____

_____

_____

_____

Cooperative-Learning Activity
1.4 Diagonals of *n*-gons

Group members: 4

Materials: paper and a pencil

Responsibilities: Draw polygons with diagonals and check partners' drawings.

Preparation: Polygons are closed figures formed by three or more segments that intersect two other segments at a vertex. A diagonal is a segment that joins two nonadjacent vertices and passes through the interior of the polygon. Can you find the number of possible diagonals for an *n*-sided polygon (an *n*-gon)?

Procedures: Use the following procedures to find the number of diagonals for an *n*-gon:

1. Each member selects a polygon and draws it on his or her own paper. They then draw diagonals in their polygon to find the maximum number of diagonals possible.

2. Each member exchanges his or her drawing with another group member to check one another's work.

3. Repeat the procedure of drawing a polygon with diagonals and recording your partner's results in the table.

Polygon	Number of sides	Process	Number of diagonals
triangle	3	$\frac{3(3-3)}{2} = \frac{0}{3} = 0$	0
quadrilateral	4		
pentagon			
hexagon			
heptagon			
octagon			
nonagon			
decagon			

4. The expression for the maximum number of diagonals of a triangle is $\frac{3(3-3)}{2} = \frac{0}{3}$. As a group, experiment by altering this expression so that it gives the maximum number of diagonals for each polygon in the table. Record the group's answers in the process column.

5. Study your completed process column and write the variable expression for the maximum number of diagonals for an *n*-gon.

Cooperative-Learning Activity
1.5 Factors and Divisibility Game

Group members: 4

Materials: spinner, 2 numbered cubes, paper, pencil, and a calculator

Roles: **Roller** rolls number cubes that are used to make two 2-digit numbers

Reader reads aloud the rolled numbers and makes two number pairs

Recorder records all numbers and keeps running totals of each member's score

Checker uses a calculator to verify the divisibility of any member's questionable answers

Preparation: Use a paper clip as a pointer by placing a pencil point through the paper clip at the center of the spinner. Spin the paper clip and observe where it lands. Spin again if the paper clip lands between two sections.

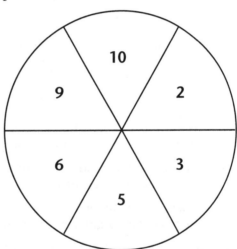

How to Play: 1. The players decide on their roles. Taking turns, each player spins the spinner to determine his or her divisibility number for the round. Each player's number must be unique, and players must spin again until they get a number that has not already been selected. The Recorder records each player's divisibility number.

2. The Roller rolls both number cubes. The Reader studies the numbers on the number cubes and makes two numbers from the numbers rolled. For instance, if a player rolls a 2 and a 1, the Reader makes the numbers 21 and 12.

3. Each player determines if their divisibility number can evenly divide either or both of the stated numbers. Each player scores 4 points if both stated numbers are divisible by his or her divisibility number, 2 points if one stated number is divisible, and 0 if neither stated number is divisible.

4. The Checker uses a calculator to verify any questionable results.

5. The Recorder records the score that each member receives and gives all players their current total of points. The first player to reach 15 points is the winner.

6. Rotate the roles until each player has been the Roller. A new round begins when each player has been the Roller. At the beginning of each new round, each player spins the spinner again to generate new divisibility numbers.

Cooperative-Learning Activity
1.6 Bean Count

Group Members: 3

Materials: dried beans and a pencil

Roles: **Counter** adds beans to a pile to represent the number of beans each day

Recorder writes down the number of beans on the table

Organizer gathers and returns the materials

Preparation: As a group, you will solve a problem by using several problem-solving strategies.

Procedures:

1. The Organizer gathers the materials for the activity.

2. The Counter places 4 beans in the center of the table to represent the number of beans starting on day 1. The Recorder completes the table for day 1.

Day	Number of beans
1	
2	
3	

3. As a group, calculate how many beans should be added on day 2 so that the total number of beans in the pile is 4 times as many as on day 1.

4. The Counter then adds that number of beans to the day 2 pile, and the Recorder completes the table for day 2.

5. As a group, complete the questions below, with the Recorder completing the table. The Organizer is responsible for returning the materials.

   a. As a group, decide how many beans are on day 3. _____

   b. As a group, describe another way to find the answer to the problem for day 2.

   _____

   c. Use your method to find how many beans there will be after 10 days and after n days.

   _____

   d. Use the guess-and-check method to calculate when there will be at least 65,000 beans.

   _____

   e. As a group, discuss which method works best in calculating these amounts. Discuss how exponents and factors play a role in your calculations.

   _____

Cooperative-Learning Activity
1.7 Order of Operations Game

Group Members: 4

Materials: scissors, cup or envelope, numbered cube, paper, pencil, and a calculator

Roles: **Roller** selects a folded slip of paper and rolls number cube

Reader reads aloud the expression selected and makes sure each member can see the expression

Recorder writes down values of variables and keeps a running total of each member's score

Checker uses a calculator to evaluate expressions and determine if each member's answers are correct

Preparation: Cut out the expressions in the table, fold them in half so that the equations are hidden, and place them in a cup or envelope.

$3y + x + z$	$x^2 + y^2 + z^2$	$2x + y \cdot z$
$x(y + x) + x$	$x + z + z \cdot y$	$4x + 3y \cdot z$
$\dfrac{x^2 + z}{y + x}$	$\dfrac{3z + y}{2x}$	$\dfrac{xy}{yz^2}$

How to Play:

1. The Roller first selects one folded paper from the cup and gives it to the Reader, who reads the expression and makes sure every group member can examine the selected slip of paper.

2. The Roller then rolls a number cube three times to determine the values for x, y, and z. The Recorder writes the values for each variable after each roll.

3. Each member evaluates the selected expression with the values determined for x, y, and z. Round answer to the nearest hundredth when necessary.

4. The Checker uses a calculator to check the solution and determines who evaluated the expression correctly. The Recorder records the score that each member receives.

   - The Roller receives 2 points.

   - No points are awarded for wrong answers.

   - The winner is the first person who gets at least 10 points.

5. For the next round, the roles are rotated: the Roller becomes the Reader, the Reader becomes the Recorder, the Recorder becomes the Checker, and the Checker becomes the Roller.

NAME _____ CLASS _____ DATE _____

Cooperative-Learning Activity
1.8 Property Possibilities

Group Members: 3–5

Materials: 3 number cubes and a pencil

Roles: **Roller** rolls a number cube to get three numbers that are used to make an equation

Equation Makers make a possible equation that has the same solution

Preparation: Use the following example as a guide for beginning the activity. You may first want to read through the procedures.

The Roller rolls the numbers 1, 2, and 4 and writes an equation with a solution of 6.

The Equation Makers come up with the following equations and properties:

$1(2 + 4) = 6$	Equation
$1(2) + 1(4) = 6$	Distributive Property
$2 + 4 = 6$ and $4 + 2 = 6$	Commutative Property of Addition

Procedures: 1. The Roller rolls three number cubes to obtain three numbers. The Roller then selects one of the properties covered in Chapter 1 to solve an equation with three variables. The Roller tells only the solution of the equation to the group.

2. The Equation Makers try to find as many ways to get the same solution as the Roller. The Equation Makers should take turns sharing their work and telling which property they used to solve the problem.

3. One of the Equation Makers should act as a recorder and write each unique solution in the table.

4. Rotate the role of Roller and repeat the procedures. The activity is finished when every member has been the Roller.

Numbers chosen	Combination possibilities

HRW material copyrighted under notice appearing earlier in this work.

Cooperative-Learning Activity
2.1 Equation Spin-Off

Group Members: 2

Materials: paper clip, two different colored pencils, game board, and paper or a paper plate

Responsibilities: Play a game, taking turns using a spinner and recording outcomes on a game board.

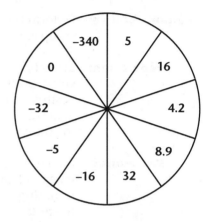

Preparation: To use the spinner, hold a pencil through one end of a paper clip and place the pencil point at the center of the spinner. Then spin the paper clip around the pencil like an arrow. Spin again if the paper clip lands between two numbers.

How to Play:
1. One player spins the spinner and, using a colored pencil, enters the outcome of the spin in one space on the game board to make a true equation.

2. The other player checks to make sure that the equation formed by his or her partner is a true equation.

3. Switch roles after every spin whether or not an equation can be completed. Continue completing equations until all of the spaces are filled in on the game board.

4. The game concludes when all of the spaces are full. The player who filled in the most spaces correctly is the winner.

Game Board

_____ = |5|	_____ = |−16|	_____ = |4.2|
_____ = 32° above zero	_____ = |−8.9|	_____ = a credit of $4.20
_____ = 16 years ago	_____ = a loss of 5 yards	_____ = 32° below zero
_____ = |0|	_____ = |16|	_____ = a credit of $8.90
_____ = a depth of 340 meters	_____ = the opposite of 0	_____ = |−5|

Cooperative-Learning Activity
2.2 Balance It Out

Group Members: 2–4

Materials: pencil

Responsibilities: Perform the calculation for each transaction, compare results, and record a group answer.

Preparation: Many banks handle credits and debits as positive and negative numbers. When you deposit money into an account, it may appear on your bank statement as a positive number. When you write checks or withdraw funds, it may show up on your bank statement as a negative number. Computers perform calculations on the transactions and come up with an ending balance.

Procedures: Each group member should perform the calculations on the first transaction independently. Then verify your answer with each member of your group before recording your group's response in the blanks provided. Take turns recording the results. As a group, answer the question after you complete the table.

Date	Check No.	Description	Amount	Balance
Starting Balance			$1000.00	
8-1	1001	Groceries	−$25.00	$975.00
8-1	1002	Groceries	−$45.00	_____
8-1	1003	Dentist	−$10.00	_____
8-3		ATM Cash Withdrawal	−$50.00	_____
8-4		Deposit	+$350.00	_____
8-4	1008	Department Store	−$112.00	_____
8-8		Deposit	+$200.00	_____
8-10	1010	Gasoline	−$30.00	_____
8-15	1009	Groceries	−$168.00	_____
8-15	1007	Book Store	−$59.00	_____
8-16		ATM Cash Withdrawal	−$90.00	_____
8-18	1011	Insurance	−$235.00	_____
8-18	1012	Car Payment	−$345.00	_____
8-23	1015	Mortgage	−$700.00	_____
8-24		Deposit	+$500.00	_____
8-25	1020	Groceries	−$150.00	_____
8-29	1021	Medical	− $50.00	_____
8-30		ATM Cash Withdrawal	− $20.00	_____
Ending Balance				$_____

1. What does the ending balance tell you? _____

Cooperative-Learning Activity
2.3 Comparing Integers Game

Group Members: 3

Materials: 75 index cards and a pencil

Responsibilities: Number the index cards and play a game that involves comparing three integers at a time. The largest integer wins the round.

Preparation: As a group, write the numbers from −40 to +34 on the 75 index cards. There should be one number on each card.

How to Play:
1. One player shuffles the index cards and gives each player 25 cards.

2. Each player turns over the top card on his or her pile of cards.

3. Examine each player's card. The player with the largest number on their index card wins the round and takes all three cards. These cards are grouped together, turned facedown, and placed underneath the other cards in the winner's pile. See the example below.

4. Repeat this activity until one group member has all of the cards. You may wish to set a time limit for the activity. If no one has all of the cards by the end of the time limit, the player with the most cards wins.

Examples:

| -1 | -6 | -8 |

The players have the cards shown. The player who has the index card with −1 takes all three cards because $-1 > -6 > -8$.

| -40 | 15 | 0 |

The player who has the index card with 15 takes all three cards because $15 > 0 > -40$.

| 2 | -4 | 20 |

The player who has the index card with 20 takes all three cards because $20 > 2 > -4$.

Cooperative-Learning Activity
2.4 Subtraction Tic-Tac-Toe

Group Members: 2–4

Materials: colored pencils and a game board for each group member

Roles: **Checker** verify the accuracy of the group's calculations

Recorder ensure that each player correctly places an X on the game board

Preparation: The tables show different integers from which the players can choose from.

Table 1

22	−11
8	−4
12	−18

Table 2

32	−51
41	−15
26	−33

Procedures:
1. The first player chooses one integer from Table 1 and one integer from Table 2.

2. The player then subtracts the integer selected from Table 2 from the integer selected from Table 1. The player then finds the result on the game board below and circles it with a colored pencil. The Checker verifies the calculations performed. The Recorder ensures that the result circled by the first player is correct.

3. The next player repeats the same procedure of selecting integers, subtracting them, and circling them on the game board with their own colored pencil. The winner is the player who gets four circles in a row, either vertically, horizontally, or diagonally.

Game Board

		10			40			
	15	59	−4	52	−37	−36		
−45	41	11	−33	−44	−24	47	29	
	−18	−30	−50	33	−3	−29		
	4	45	27	37	−59	63		
	55	22	73	−19	−43	−20		
		14		−20				

Cooperative-Learning Activity
2.5 Calories in Food and Exercise

Group Members: 2–4

Materials: pencil

Responsibilites: Solve problems independently, verifying the results with your group.

Preparation: The tables show how many Calories are contained in certain foods and how many Calories are "burned" by doing specific exercises for a certain amount of time.

Food	Number of Calories
1 hamburger	380
1 bowl vegetable soup	90
1 oz skim milk	10
1 candy bar	240

Exercises	Calories burned
running	15 Calories/min
bicycling	11 Calories/min
swimming	7 Calories/min
walking	3 Calories/min

Procedures: Each member solves each problem but can ask other group members for assistance. As a group, reach a consensus on each answer before recording a group response.

1. Suppose that someone in your group swam for 1 hour, bicycled for 2 hours, and ran for $\frac{1}{2}$ hour. How many Calories would they expend? _____

2. Suppose that this group member wanted to regain the Calories expended by doing the activities in Exercise 1. About how many candy bars would he or she have to eat to regain the calories? _____

3. Suppose that a group member eats 1 hamburger and a bowl of vegetable soup and drinks 8 oz of skim milk for lunch. He or she also eats a candy bar. How many Calories are gained? _____

4. Calculate, to the nearest minute, how long this group member would need to bicycle in order to use all of the Calories consumed for lunch. Round your answer, if necessary. _____

5. How long would this group member have to walk in order to expend the Calories gained from this meal? _____

6. How long would this group member have to run in order to expend the Calories gained from this meal? _____

Cooperative-Learning Activity
2.6 Describing Data in a Table

Group Members: 4

Materials: pencil

Responsibilities: Each group member writes a linear equation and graphs data from a table.

Preparation: Each member selects one of the four data tables below.

Procedures:
1. Each member generates a linear equation to describe the data in their table.
2. Each member graphs the data in his or her table on graph paper.
3. Then each member completes his or her data table by using the graph.
4. Each member exchanges his or her graph with another member, who checks the work.
5. As a group, record the verified results in the space provided for each data set.
6. What did group members learn from comparing their results with the results they checked? _____

A.

x	y
0	−5
1	−4
2	−3
3	−2
4	
5	

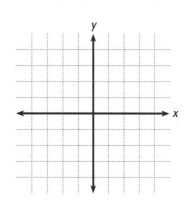

_____

B.

x	y
0	10
1	13
2	16
3	19
4	
5	

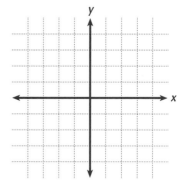

_____

C.

x	y
−2	6
−1	4
0	2
1	0
2	
3	

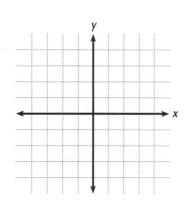

_____

D.

x	y
−2	8
−1	4
1	−4
2	
3	

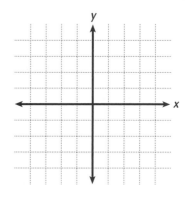

_____

Cooperative-Learning Activity
2.7 Decipher the Code

Group Members: 2

Materials: pencil

Roles: **Difference Finder** finds differences for the given sequence until the differences are constant

Letter Remover crosses out letters according to the constant difference

Preparation: Use the example as a guide for the procedure of the activity.

Procedures: 1. The Difference Finder finds the first differences for the first sequence. If the differences are not constant, the Difference Finder finds the second differences. Differences are found until a constant differences occur.

2. The Letter Remover uses the value of the constant difference to cross out the unwanted letters in the group of letters shown for the sequence.

3. As a group, combine the remaining letters and spell out the name of a famous mathematician.

4. Repeat the procedure for the next problem but rotate your roles.

* *

Example Sequence: 0, 3, 8, 15 . . .

Letters: A B̸ C D̸ E F̸ G
1 2 1 2 1 2 1

Differences: 0   3   8   15

3   5   7 ← **1st Differences**

2   2 ← **2nd Differences**

Cross out every second letter since the constant is 2. Now you are left with ACEG.

* *

1. Sequence: 1, 2, 5, 10, 17, . . .   Letters: E A U L C E L T I T D E

Constant: _____

2. Sequence: 16, 19, 22, 25, . . .   Letters: P Y H T H N A G M O R A A S T

Constant: _____

3. Sequence: 2, 4, 10, 20, 34, . . .   Letters: D E S M C A R D T E S T

Constant: _____

4. Sequence: 10, 11, 14, 19, . . .   Letters: E M U S L A E T R

Constant: _____

5. As a group, make up your own sequence and letters for a mathematical term instead of a famous mathematician. Trade your problem with another group.

Cooperative-Learning Activity
3.1 Comparing Fractions Game

Group members: 2

Materials: 20 index cards and a calculator

Roles: **Dealer** deals the cards

Writer writes a summary of the methods used in the activity

Preparation: Students will play a game that involves comparing two numbers at a time from the following list:

$$\frac{1}{2} \quad \frac{6}{12} \quad \frac{10}{20} \quad \frac{1}{4} \quad \frac{3}{4} \quad \frac{2}{3} \quad 1 \quad \frac{1}{5} \quad \frac{2}{6} \quad \frac{4}{5} \quad \frac{2}{5} \quad 0 \quad \frac{6}{8} \quad \frac{2}{10} \quad \frac{5}{6} \quad \frac{5}{7} \quad \frac{6}{7} \quad \frac{3}{12} \quad \frac{4}{6} \quad \frac{8}{11}$$

Note: When two fractions have the same denominator, the fraction with the larger numerator is the larger of the two fractions.

Procedures:

1. Each player takes 10 index cards. One player writes the first 10 numbers on his or her index cards and the other player writes the last 10 numbers on his or her index cards.

2. The players designate a Dealer. The Dealer then shuffles all 20 index cards and deals 10 cards facedown to each player. Each player puts his or her cards facedown in a pile.

3. The players each take one card from the top of their pile and place it faceup in the center of the playing area. The players decide whose card shows the larger number by renaming any fractions with common denominators. The player with the larger number wins both cards and puts them in a separate pile.

   If the two cards show equivalent numbers, the players leave the cards in the center of the playing area and play two more cards. The players then compare the numbers on the two new cards, and the player who wins that round wins all of the cards in the center. Whenever a tie results, this procedure is followed until one player wins the round.

4. When the players run out of cards to play, they continue playing with the stack of cards they won in previous rounds.

5. The first player to get all of the cards wins.

6. The player who was not the Dealer is the Writer of the group. The Dealer helps the Writer write an explanation of how they determined the winner of each round. The Writer reads the explanation to the class.

   _____

   _____

   _____

Cooperative-Learning Activity
3.2 Finding the Least Common Multiple

Group members: 3

Materials: regular six-sided number cube

Roles: **Roller** rolls the number cube

Recorder records the two numbers designated by two rolls of the number cube

LCM Finder finds the LCM of the recorded numbers

Preparation: The least common multiple, or LCM, of a pair of numbers is the smallest number that is a whole-number multiple of both numbers. The LCM of the denominators of two fractions is the least common denominator of the fractions.

Procedure: 1. The group decides the role of each member. The Roller rolls the number cube twice.

2. The Recorder records the number on the first roll of the cube and the sum of the numbers of the two rolls of the cube in the table below.

3. The LCM Finder finds the least common multiple of the two numbers recorded in the table. The other members verify that the LCM is correct, and the Recorder records the correct LCM in the table.

4. Repeat Procedures 1–3, switching roles until each member has had 3 turns in each of the roles.

5. The group discusses any pattern it notices and designs a set of rules for finding the least common denominator of two fractions. Then each member uses the rules to rename the denominators of $\frac{11}{12}, \frac{5}{16},$ and $\frac{7}{24}$ with the LCM.

   How does renaming the denominators with the LCM make it easy to compare the fractions?

First number rolled	Sum of the two numbers rolled	LCM
_____	_____	_____
_____	_____	_____
_____	_____	_____
_____	_____	_____
_____	_____	_____
_____	_____	_____
_____	_____	_____

Cooperative-Learning Activity
3.3 Concentration: A Decimal Equivalent Game

Group members: 3–4

Materials: plain white paper, construction paper, pencil, scissors, and calculator

Responsibilities: Find equivalent fraction and decimal pairs.

Preparation: You will need to find the decimal equivalents for some of the fractions listed below.

$\frac{1}{3}$ $4\frac{2}{3}$ $\frac{1}{8}$ $3\frac{3}{5}$ $\frac{4}{9}$ $16\frac{9}{25}$ $2\frac{1}{5}$ $\frac{17}{20}$ $\frac{2}{3}$ $3\frac{2}{25}$ $\frac{7}{10}$ $\frac{3}{4}$ $\frac{3}{5}$ $1\frac{1}{3}$ $10\frac{1}{10}$ $\frac{13}{20}$

Since the fraction bar indicates division, the decimal equivalent for each of these fractions can be found on a calculator by dividing the numerator by the denominator. When a fraction is equivalent to a repeating decimal, the repeating number(s) will completely fill the calculator display.

Procedure:

1. The group chooses one member to draw an 8-inch-by-8-inch game board on a plain white piece of paper, which is then divided into 16 equal sections.

2. Each player cuts out at least four 1.5-inch-by-1.5-inch squares from a piece of construction paper. A total of 16 squares must be cut out for each group.

3. The players take turns choosing a fraction from the list above until the group has chosen a total of 8 fractions. As a player chooses a fraction, he or she writes the fraction on one of the squares of construction paper. Then the player determines the decimal equivalent for the fraction. The other players must verify the decimal. Calculators may be used. Once the equivalent decimal is verified, the player writes that decimal on another square of construction paper.

4. The players place all of their squares of construction paper facedown. One player is chosen to mix up the 16 squares.

5. The players take turns placing the squares facedown on the game board so that no player knows the location of any fraction or decimal.

6. The object of the game is to uncover the largest number of fraction and decimal pairs. The youngest player begins the game by choosing two squares and turning them faceup. If the numbers on the squares are equivalent, the player keeps the pair of squares; if the numbers are not equivalent, the player turns the squares facedown again on the game board.

7. Repeat Procedure 6 until each of the remaining players has had a turn. The players should take their turns in a clockwise direction beginning with the first player.

8. Continue the game until the whole game board is uncovered. The player with the most pairs wins.

9. The group identifies the fractions uncovered that are repeating decimals. What are the denominators of these fractions? Each member chooses two more fractions with one of these denominators and uses a calculator to find the equivalent decimal. The members compare their results. One member is chosen to write a summary of the results with input from the other members.

Cooperative-Learning Activity
3.4 Exploring the Stock Market

Group members: 2

Materials: newspaper with a stock market report and a pencil

Responsibilities: Find the previous day's closing price of several stocks.

Preparation: Companies sell parts or *shares* of ownership in their company to raise money for their businesses. These ownership shares are called *stocks*. The people who buy the stocks are called *stockholders*. For a variety of reasons, stockholders often want to sell their stock or buy more stock. In places called stock exchanges, stocks are traded in eighths of dollars. The financial sections of large newspapers list the last price at which stocks were sold that day on the stock exchange. This price is called the *closing price*. The stock report also lists the change between the current day's closing stock price and the previous day's closing price.

Procedure:
1. Each member chooses five stocks from the financial section of a newspaper.

2. Each member reads across the columns of stock prices and finds the current day's closing price for each stock that he or she has chosen. The members record the closing prices in the table below.

3. Each member reads across the columns of stock prices and finds the change in closing price between the previous day and the current day. The members record the changes in the table.

4. Each member uses the results in the table to find the previous day's closing price for each of his or her stocks by adding or subtracting the current closing price and the change. If the change is positive, subtract the change from the current closing price. If the change is negative, add the absolute value of the change to the closing price.

5. The members check each other's calculations and discuss any discrepancies.

Name of stock	Closing price per share	Change	Previous day's closing price per share
1.			
2.			
3.			
4.			
5.			

6. The members discuss how adding and subtracting fractions play an important role in the stock market.

7. The members review their list of stocks. Would you purchase any of these stocks? Which ones? Why?

Cooperative-Learning Activity
3.5 Adjusting Recipes

Group members: 2–4

Materials: cookbook or recipes and a pencil

Responsibilities: Increase and decrease the amounts of the ingredients in a recipe.

Preparation: Suppose that the student council has asked your group to provide the food for an upcoming school party. Since the council members are unsure about how many students will actually attend, they have asked you to work up several scenarios so that they can see how much the food will cost.

Procedures:

1. The members decide on an appetizer recipe that they think the guests would enjoy. One member is chosen to record the ingredients and their amounts in the first two columns of the table below. The other members should verify that the ingredients were recorded correctly.

2. Each member doubles the ingredients in the recipe.

3. The members compare their answers, making any necessary corrections to their own answers. A member other than the member chosen in Procedure 1 is chosen to record the correct answers in the table.

4. Repeat Procedures 2 and 3 two more times, to triple the recipe and then to quadruple it. All members should have a turn recording the results in the table.

5. The group estimates the number of people that each recipe will serve.

Ingredients	Amounts			
	Original recipe	Doubled recipe	Tripled recipe	Quadrupled recipe
Number of servings				

6. The group shares their results with the class. One member tells what the recipe is and what ingredients are used. The other members of the group take turns presenting to the class the amounts needed to double, triple, and quadruple the recipe.

7. Suppose that you wanted to prepare the recipe for a smaller group. Adjust the original recipe by half and then by one-quarter. Show the amounts of each ingredient that you will need.

Cooperative-Learning Activity
3.6 Scale Drawings

Group members: 2–4

Materials: large sheet of plain paper, pencil, and ruler

Responsibilities: Make a scale drawing of a room in your school.

Preparation: Your class has been asked to make a scale drawing of a room in your school. Making the scale drawing involves choosing a scale to use in making a reduced drawing of the room. For example, if the scale used is 1 inch = 3 feet, then every inch on the drawing represents 3 feet in the actual room. This scale, which expresses the ratio $\dfrac{\text{length in drawing}}{\text{actual length}}$, can be written as $\dfrac{1 \text{ inch}}{3 \text{ feet}}$. If a room is 30 feet long, on a scale drawing with this scale, the room length would be 10 inches because $\dfrac{1}{3} = \dfrac{10}{30}$.

Procedures:

1. Each group should choose or be assigned a room in the school, such as the library, the gym, a classroom, or the cafeteria, for which the group will make a scale drawing.

2. The group measures the length and width of the room, rounding each length or width to the nearest unit or half unit. In a group of four, the members choose teams of two. One team measures the length of the room and the other team measures the width. If the group has two members, one member measures the length, and the other member measures the width. The groups record their measurements in the table below. The two teams or the two members check each other's measurements and make adjustments, if necessary.

   If the group has three members, two members should be chosen to measure the length and width, which are then recorded. The third member checks the measurements and makes any necessary adjustments.

3. Each member converts the actual measurements to scale measurements by using a proportion that involves the scale.

4. The members compare their answers and record the correct scale measurements in the table.

5. Each member uses the scale measurements to make a scale drawing of the group's room.

6. The members compare their drawings. One member volunteers to present his or her drawing to the class, explaining the scale that was used.

7. The group discusses how they used proportions to make their scale drawings. What are some advantages to making a scale drawing?

Room:		
	Actual measurement	Scale measurement
Length		
Width		

Cooperative-Learning Activity
3.7 Classroom Statistics

Group members: 4–6

Materials: paper, pencil, and calculator

Responsibilities: Collect data and express the data as a ratio, a decimal, and a percent.

Preparation: Suppose that the school newspaper is planning to publish a report titled "Facts and Figures About Algebra Students." The newspaper staff has asked algebra students to gather data about some characteristics of the members of their class. As a group, you will gather this information and express your data as a percent for the newspaper article.

Procedures:

1. Each member is assigned one of the characteristics listed in the first column of the table below. The members collect the data for their characteristic from the members in their group.

2. The members take turns recording the data they collected in Procedure 1 in the second column of the table below. In the third column of the table, the members record the ratio of $\dfrac{\text{number of students with characteristic}}{\text{total number of students in group}}$ for the data they collected.

3. Each member calculates the decimal value and the percent for each ratio in the table. Calculators may be used. The members compare their answers and discuss any discrepancies.

4. Each member records the correct decimal value and percent for his or her characteristic in the table.

Characteristic	Number of students	Number of students with characteristic / Total number of students in group	Decimal value	Percent
1. Hair color • red • blonde • brunette • other				
2. Eye color • blue • brown • green • other				
3. Sex • male • female				

5. The groups pool their data. What is the new denominator for each ratio? How do you find the numerator for each ratio? Each group compares its data with the pooled data. What does it mean if any of your group's percents are close to the class percents? What does it mean if they are different?

Cooperative-Learning Activity
3.8 Finding an Experimental Probability

Group members: 2

Materials: a pair of regular number cubes

Responsibilities: Conduct a probability experiment and find the experimental probability.

Preparation: With a partner, students will conduct their own simulation to find the experimental probability that the sum of two numbers rolled on a single roll of a pair of number cubes is greater than 7. A sum greater than 7 is a successful event in this simulation.

Procedures: 1. Each member begins by estimating the experimental probability of rolling a sum greater than 7 on 50 rolls of a pair of regular six-sided number cubes.

2. The members take turns rolling the cubes. Each time a member rolls the cubes, he or she records the numbers rolled and their sum in the table below. Complete this activity 50 times.

Trial	Numbers	Sum	Trial	Numbers	Sum	Trial	Numbers	Sum	Trial	Numbers	Sum
1.			14.			27.			40.		
2.			15.			28.			41.		
3.			16.			29.			42.		
4.			17.			30.			43.		
5.			18.			31.			44.		
6.			19.			32.			45.		
7.			20.			33.			46.		
8.			21.			34.			47.		
9.			22.			35.			48.		
10.			23.			36.			49.		
11.			24.			37.			50.		
12.			25.			38.					
13.			26.			39.					

3. Each member should count and verify the number of successful events.

4. Each member finds the experimental probability by writing a ratio that compares the number of successful events to the number of trials. The members decide whether the experimental probability is close to the estimate made in Procedure 1.

5. The groups pool the results of their simulations. What is the experimental probability of the pooled results? The members discuss whether they expect this probability to be close to the experimental probability of their group's simulation.

Cooperative-Learning Activity
3.9 Designing an Experiment

Group members: 2

Materials: number cubes

Responsibilities: Determine the experimental probability of an experiment based on the theoretical probability of the experiment.

Preparation: There are six possible outcomes for the roll of a regular number cube, so there are 6 × 6, or 36, possible outcomes from rolling two number cubes together. The grid below shows the 36 sums that are possible when two number cubes are rolled together.

Possible Sums of Two Rolls

			Second cube				
		1	2	3	4	5	6
	1	2	3	4	5	6	7
	2	3	4	5	6	7	8
First	3	4	5	6	7	8	9
cube	4	5	6	7	8	9	10
	5	6	7	8	9	10	11
	6	7	8	9	10	11	12

Procedures:
1. The members discuss, then answer the following questions:
   - How many possible sums are there?
   - How many sums are greater than 7?

2. The members write a reduced ratio that compares the number of sums greater than 7 with the total number of possible sums. This is the theoretical probability of rolling a sum greater than 7 when two number cubes are rolled together.

3. Each member uses the theoretical probability found in Procedure 2 to estimate the number of sums that would be greater than 7 if the number cubes were rolled 100 times. The members compare their answers.

4. The members discuss what they think the theoretical probability of rolling a sum greater than 7 on 1000 rolls of the number cubes is. How is this related to the experimental probability of the experiment?

5. If the class completed the Cooperative-Learning Activity for Lesson 3.8, the members should compare the results of the experiment they performed in that activity with the theoretical probability found in Procedure 2 of this activity. Are the probabilities nearly equal?

6. The members compare the theoretical probability with the experimental probability. How are they similar? How are they different? One member writes a summary of the group's discussion.

Cooperative-Learning Activity
4.1 Creating Lines and Angles With Paper Folds

Group members: 4

Materials: plain sheets of white paper and a protractor

Roles: **Folder** folds a piece of paper to create angles or lines

Writer writes an explanation of how the angle or lines were created by the paper folding

Checker checks the results of the paper folding by using a protractor

Spokesperson explains to the class how the group created each angle and set of lines

Preparation: The group must find a way to fold sheets of paper to show the following lines and angles:

- right angle and perpendicular lines
- parallel lines
- a 45° angle
- a 135° angle
- a 225° angle

Procedures: 1. The Folder folds a piece of paper so that a right angle and perpendicular lines are formed by the folds. If the Folder is having difficulty, the other members can offer suggestions.

2. When the Folder has completed the activity, the Checker should check the results by using a protractor.

Use these steps to measure an angle with a protractor.

- Place the center of the protractor's straightedge at the vertex of the angle.

- The straightedge of the protractor should lie along one side of the angle.

- Read the number of degrees at the point where the other side of the angle falls on your protractor.

Members should discuss how they know which scale on the protractor to use when measuring an angle.

3. With input from the other members, the Writer writes an explanation of how the activity was completed.

4. Members repeat Procedures 1–3 for the remaining lines and angles. Switch roles so that different Folders and Checkers can participate in these activities. Be sure to consider how you can use a protractor to verify that folded lines are parallel. Can you use any completed angles to create the 135° and 225° angles?

5. The Spokesperson for each group makes a presentation to the class, explaining how the folds were made. How are the groups' results alike? How are they different? How can more than one fold show the same lines or angles?

Cooperative-Learning Activity
4.2 Finding Supplements and Complements of Angles

Group members: entire class

Materials: pencil, protractor, and models of angles

Responsibilities: Determine the supplement or complement of an angle.

Preparation: At least 12 different pairs of supplementary and complementary angles that have been cut out of paper or cardboard by the teacher. Each student will receive an angle model. Every student's angle will be the complement or supplement of another student's angle.

Procedures:

1. Mix up all of the angle models.

2. Each member of the class chooses an angle and then finds someone in the room with the complement or supplement of his or her angle. You can do this by seeing if the two angles form a straight line, or if the two angles form a right angle.

3. After each member has found a complementary or supplementary angle, measure the angles with a protractor to verify your results. Write the degree measure on one side of the angle. Record the class results in the space provided.

Angle measure	Angle measure	Supplementary or complementary?	Angle measure	Angle measure	Supplementary or complementary?

4. Repeat the activity after placing the angles with their measures face down. This time, after each member has chosen an angle, each member announces the kind of angle—acute, obtuse, or right—that he or she has. Other members then decide whether their angle is a possible supplement or complement of the announced angle. The announcer reveals the measure of his or her angle and any angles that are complementary or supplementary to it.

5. What kind of angles form a complementary pair? What kind of angle is the supplement of an acute angle? of an obtuse angle? of a right angle?

_____

_____

_____

Cooperative-Learning Activity
4.3 Finding Measures of Angles Formed by a Transversal

Group members: 4

Materials: construction paper, scissors, pencil, and protractor

Responsibilities: Measure the angles formed when two parallel lines are cut by a transversal.

Preparation: Students are shown how to fold a piece of construction paper in thirds. The folds of the paper should model two parallel lines.

Procedures:

1. Each member folds a piece of construction paper in thirds to model a pair of parallel lines and then cuts the paper in half along a diagonal to model a transversal.

2. Begin by identifying a pair of alternate interior angles formed by the folds and the transversal. One member manipulates the two halves of his or her paper so that the group can make a conjecture about the measures of these angles. Check the conjecture by examining another pair of alternate interior angles. Another member should record the conjecture.

   _____

3. Repeat Procedure 2 for alternate exterior angles, corresponding angles, and interior angles on the same side of the transversal. Each member should have a turn manipulating his or her two halves of paper.

   _____

   _____

4. Each member uses a protractor to measure the angles identified in Procedures 2 and 3.

5. Compare each member's measurements to make sure the answers match.

6. Modify your list of conjectures based on any other information you found after measuring the angles with your protractor. (Hint: Add pairs of angle measures to find any patterns you may have missed.)

7. Write a group summary of the conjectures.

   _____

   _____

   _____

   _____

Cooperative-Learning Activity
4.4 Drawing A Regular Tessellation

Group members: 2–3

Materials: paper, construction paper, pencil, and models of regular polygons

Responsibilities: Draw a regular tessellation.

Preparation: For this activity, each group will use models of regular polygons having 3, 4, 5, 6, 8, and 10 sides that have been cut out of cardboard or manila folders.

When polygons are fitted together to cover a plane with no space between the polygons and no overlapping, the pattern is called a tessellation. A regular tessellation is made up entirely of congruent regular polygons which meet in such a way that no vertex of one polygon lies on a side of another polygon.

Procedures: 1.   Each member chooses two or three of the regular polygon models and studies or manipulates them to see which polygons he or she believes will form a regular tessellation. The polygons should be divided equally among the members.

2.   The members take turns explaining to the group why they think their polygons can form a regular tessellation. The members then place the polygons that will form a regular tessellation in one pile.

3.   One member is chosen to equally distribute the polygons that will form a regular tessellation to the group members. Each member then uses the construction paper and his or her polygon(s) to draw a regular tessellation.

4.   The group chooses a member to write a paragraph outlining the group's results with input from the group. How many regular tessellations did you find? Justify your response with information from the activity.

_____

_____

_____

_____

_____

_____

_____

5.   A reader is chosen by each group to read the group's paragraph aloud to the class. Any differences in the groups' results should be discussed by the class.

Cooperative-Learning Activity
4.5 Area and Perimeter Scavenger Hunt

Group members: 3–4

Materials: pencil, calculator, ruler, and yardstick or measuring tape

Roles: **Measurer** measures the length and width of the rectangular surfaces

Calculator calculates the perimeter and area of the rectangular surfaces

Scorer finds the group's score

Preparation: The groups will have to find the perimeters and areas of several rectangular surfaces in their classroom or in other parts of the school. These rectangles and their location are written on the chalkboard or on an overhead. The unit of measure to be used for each object is also indicated.

Procedures:
1. All players in a group will search for rectangular objects.

2. When the group finds an object, the Measurer measures the dimensions of the rectangle. From these dimensions, the Calculator calculates and records the area and perimeter of the rectangle. If the group has four players, two players can be chosen as Calculators, with one Calculator finding the perimeter of the rectangle and the other Calculator finding the area of the rectangle. All players should guide the Measurer and the Calculator, if necessary.

3. The teacher records the perimeters and areas of the objects on the chalkboard or on an overhead. The Scorer compares the team's perimeter and area measurements with those recorded by the teacher. For each team measurement that matches the teacher's measurement, the Scorer gives the team 3 points. For each team measurement that *does not match* the teacher's measurement, the Scorer gives the team a score of -1.

4. The Scorer tallies the team's score. The other players verify the score. The team with the greatest point total wins the scavenger hunt.

5. To summarize the activity, each team chooses a different object from the hunt and discusses these questions:

   • What is the greatest possible area of a rectangle with the perimeter of your object?

   • What shape would this area have?

   • What are the dimensions of this figure?

   Discuss any advantage to maximizing the area of your rectangular surface. Choose a member to report your findings to the class.

   _____

   _____

   _____

   _____

Cooperative-Learning Activity
4.6 Finding Areas of Polygons on a Geoboard

Group members: 5

Materials: pencil, calculator, ruler, geoboard, and rubber bands

Roles: **Director** assigns roles to the members of the group and directs the activity

Creator creates a polygon on the geoboard

Estimator estimates the area of the polygon

Reader states the formula for the area of the polygon modeled and the dimensions of the polygon

Checker calculates the actual area of the polygon

Preparation: Polygons can be modeled by using a geoboard. The pegs on the geoboard can represent the vertices of the polygon. A rubber band stretched between two pegs can represent one side of a polygon. Use the geoboard to find the area of several triangles, parallelograms, and trapezoids.

Procedures:
1. The members choose a Director for the group. The Director assigns the other four roles to the rest of the members.

2. The Creator creates either a triangle, a parallelogram, or a trapezoid on the geoboard. Then the Director asks the Estimator to estimate the area of the figure.

3. The Director asks the Reader to state the formula for the area of the modeled polygon and the dimensions of the polygon. The Director then asks the Checker to check the reasonableness of the estimate by calculating the actual area of the figure. The Director leads the group in a discussion of any discrepancies between the estimated area and the calculated area.

4. Repeat Procedures 1–3 with a different Director until each member has had a turn as Director. The other members should also be assigned new roles each time the procedure is repeated. It is the Directors' responsibilities to make sure that all three polygons are modeled.

5. Repeat the procedure until each member has created at least two figures.

6. All members of the group should discuss how the geoboard can help them find the area of a triangle, a parallelogram, or a trapezoid. Which figure's area is the easiest to estimate on the geoboard?

7. The group chooses one member to write a summary of their discussion.

_____

_____

_____

_____

Cooperative-Learning Activity
4.7 Creating a Radical Spiral

Group members: 2

Materials: paper, pencil, ruler, and protractor

Responsibilities: Draw a series of right triangles, find the length of the hypotenuse of each triangle, and examine the pattern that develops.

Preparation: The length of a hypotenuse can be calculated from the lengths of the legs of a right triangle. By the "Pythagorean" Right-Triangle Theorem, if c represents the length of the hypotenuse and a and b represent the lengths of the legs, then $c^2 = a^2 + b^2$. From this equation, it follows that $c = \sqrt{a^2 + b^2}$.

Procedures:

1. Each group member draws an isosceles right triangle in the middle of a piece of paper. Each leg of the triangle should be 1 inch. Use a protractor to verify that the triangle is a right triangle. Label the lengths of the legs of the triangle.

2. Each member finds the length of the hypotenuse of his or her triangle by using the "Pythagorean" Right-Triangle Theorem. The partners verify their answers, which should be kept in radical form, with each other. Then each member labels the length of the hypotenuse of his or her triangle.

3. Each member draws a new right triangle, using the hypotenuse of the just completed triangle as a leg for the new triangle. The other leg of the new triangle is drawn from the left end of that hypotenuse and is 1 inch long. Label that leg with its length. The partners compare drawings and make any adjustments to their drawings, if necessary.

4. Repeat Procedure 2 for the new hypotenuse.

5. Repeat Procedures 3 and 4 twelve more times, drawing a new right triangle each time using the hypotenuse of the just completed triangle as a leg of the new triangle, and making the other leg 1 inch long.

6. Each group examines their completed figures and discusses the patterns they see. What makes up the outside edge of your figure? What shape is your figure? If you were to continue this activity, what would the lengths of the hypotenuses of the next three triangles be? Why do you think this figure is called a "radical spiral"?

_____

_____

_____

_____

Cooperative-Learning Activity
4.8 Using Scale Factors To Create Similar Triangles

Group members: 3–5

Materials: pencil, $\frac{1}{4}$-inch graph paper, ruler, and protractor

Responsibilities: Draw similar right triangles by using a scale factor, and verify that the triangles are similar.

Preparation: The students will draw similar figures, using the right triangles below.

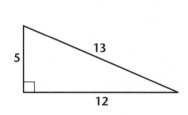 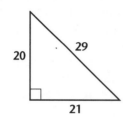 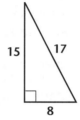 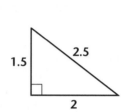 

Procedures:

1. Each member chooses one of the triangles and copies the triangle on his or her graph paper. Let each square on the graph paper represent one square unit. Each member should draw a different triangle.

2. The members exchange papers so that no members have their own drawings. The members then draw either an enlargement or a reduction of the triangle that they have just received. The enlargements and reductions should be drawn on the same piece of graph paper as the original triangle. If the smallest length on a triangle is less than 6, then an enlargement of the triangle should be drawn using a scale factor of 200%. If the smallest length on a triangle is greater than or equal to 6, a reduction of the triangle should be drawn using a scale factor of 50%. The members should draw their enlarged or reduced triangles by applying the scale factors to the legs of the original triangle.

3. The members again exchange papers. No members should have either the original triangle or the scale triangle that they drew. Each member verifies that the triangles on his or her paper are similar by measuring the acute angles of both triangles. Each member should also measure the hypotenuse of the enlarged or reduced triangle.

4. The members take turns sharing their measurements with each other.

5. The members discuss the results of the activity. Is the hypotenuse of each enlarged or reduced triangle related to the original hypotenuse by the same scale factor as the legs? How are the corresponding angles of the original triangle and the scaled triangle related? Is the shape of the original triangle preserved when the triangle is enlarged or reduced? How do you know?

_____

_____

_____

Cooperative-Learning Activity
5.1 Simplifying Algebraic Expressions

Group members: 4

Materials: map of the western United States, paper, and pencil

Roles: **Leader** confirms location of each state and asks teacher for assistance, if necessary

Recorder records answers

Coordinator makes sure all members agree on answers

Reporter summarizes the group's work

Preparation: This map represents 11 western states of the United States. An algebraic expression has been substituted for each state's name. The group will use the expressions in the exercises, which are divided into the following two parts.

Procedures: 1. The members choose the roles of Leader, Recorder, Coordinator, and Reporter. The Leader identifies the state represented by each expression.

   a. $5x - 2y$ b. $2x - 1$

2. The Coordinator makes sure that all members agree with the Leader's identification. The Recorder records the correct answer.

3. The members rotate their roles. The Leader identifies the expressions for the states, and the Coordinator makes sure that everyone agrees with the identification.

4. Each member substitutes the expressions for the states and simplifies the indicated sum or product.

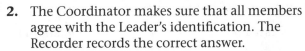

   a. Washington + Idaho   b. Nevada + Utah   c. 3(Utah) + Idaho

   d. New Mexico + Arizona   e. 2(Colorado)   f. 4(Oregon)

   g. California + Montana + Wyoming   h. 2(Arizona) + 3(Montana)

5. The members compare their answers, and the Coordinator makes sure that they agree on the answer. The Recorder records the answer.

6. The group discusses the methods they used to simplify the expressions, and the Reporter writes a summary of the methods outlined in the discussion.

7. The Coordinator confirms that the members agree with the summary.

Cooperative-Learning Activity
5.2 Evaluating and Simplifying Expressions

Group members: 4

Materials: map of the central United States, paper, and pencil

Roles: **Leader** confirms location of each state and asks teacher for assistance, if necessary

Recorder records answers

Coordinator makes sure all members agree on answers

Reporter summarizes the group's work

Preparation: This map represents 11 central states of the United States. An algebraic expression has been substituted for each state's name.

Procedures: 1. The members rotate the roles of Leader, Recorder, Coordinator, and Reporter. The Leader identifies the expressions for each state listed, and the Coordinator makes sure that everyone agrees with the identification.

2. Each member substitutes the expressions for the states. Then each member evaluates or simplifies the expressions.

   • Find the opposite of the expression for each state.

   a. Nebraska   **b.** Oklahoma

   • Simplify the algebraic expression.

   c. Missouri − Kansas   **d.** Iowa − South Dakota

   e. 2(Texas − Louisiana)   **f.** North Dakota − South Dakota

   • Evaluate the expressions for $x = 3$ and $y = -2$.

   g. Minnesota   **h.** −(Kansas)

3. The members compare their answers, and the Coordinator makes sure that they agree on the answer. The Recorder records the correct answer.

4. The group discusses whether each of the following statements is true or false:

$$\text{Texas} = -(\text{Iowa})$$
$$-(\text{Louisiana}) = -\text{South Dakota}$$

5. The group chooses a Reporter, who writes a summary of the discussion with input from the other members. The Reporter then reads his or her summary to the class.

Map expressions:
$3x$ ・ $2y - 2x$ ・ $x + 3y$ ・ $x + 3$ ・ $2y - x$ ・ $2x + y$ ・ $3x + 2y$ ・ $2y - 2$ ・ $x - 2$ ・ $x - 2y$ ・ $3x + y$

Cooperative-Learning Activity
5.3 Equation Game

Group members: 3

Materials: different colors of construction paper, scissors, pencil, paper bags, and calculator

Roles: **Leader** coordinates activity and acts as scorer

Recorder records data

Checker uses a calculator to check solutions

Preparation: The members of the group must create and then solve an addition equation during their turn of play. Algebra tiles may be used to solve the equation. The players will create their equations from two sets of squares made from construction paper. Results should be recorded in a table like this.

Round	Player	A	B	A + x = B	x	Score for round	Total score at end of round

Procedures:

1. The players cut out six 2-inch squares from one color of construction paper. The players number the squares in this set −6, −3, −1, 1, 3, and 6. These squares, which are the A squares, are placed in a bag. The players cut out a second set of six 2-inch squares from another color of construction paper. Number these squares −3, −2, −1, 1, 2, and 3. These squares, which are the B squares, are placed in another bag.

2. The players decide on their roles, which are rotated for each new round of play. To determine Player 1 for the first round, the Leader chooses a square from one bag. The other players then choose squares from the same bag. The player with the highest number is Player 1.

3. Player 1 chooses one A square and one B square and fills in the A, B, and A + x = B columns of the table, using the numbers on his or her squares. Then Player 1 solves the equation for x and writes the solution in the table.

4. The Checker substitutes Player 1's solution for x in the player's original equation and checks the solution using a calculator. If the result is B, then the player's solution is correct.

5. The Leader awards 1 point for a correct solution and subtracts 2 points for an incorrect solution. The Recorder records the scores for the round and the total scores in the table.

6. Players continue the round, playing in a clockwise direction, until all the players have had a turn.

7. Players continue playing rounds, repeating Procedures 2–6. The first player to reach 10 points wins.

8. The players take turns explaining how they solved their equations. One player is chosen to summarize the procedures that the players used to solve their equations.

Cooperative-Learning Activity
5.4 Fractional Equation Game

Group members: 4

Materials: a regular number cube, pencil, paper, and a stopwatch or a clock

Roles: **Leader** coordinates activity

Builder builds an equation

Solver solves an equation

Recorder records data

Preparation: For each round, the teacher will roll the number cube three times. The numbers showing on the cube for each roll will represent the following variables or numbers:

Roll 1		Roll 2		Roll 3	
Number	Variable represented	Number	Number represented	Number	Number represented
1	a	1	$\frac{1}{2}$	1	$\frac{3}{4}$
2	b	2	$-\frac{1}{4}$	2	$-\frac{3}{4}$
3	c	3	0.6	3	0.4
4	x	4	-0.2	4	-0.8
5	y	5	10	5	2
6	z	6	-4	6	-2

Procedures:
1. The players decide on their roles, which are rotated for each new round. The teacher rolls the cubes one at a time and records the results on the chalkboard.

2. The teacher begins timing the players for 2 minutes. In this time, the Builder uses the results of the teacher's rolls to build an addition or subtraction equation, which the Recorder writes on a piece of paper. The Solver then solves the equation.

3. When time is called, the Builder reads his or her equations and the Solver reads his or her solutions. The Recorder writes the equations and the solutions on the chalkboard.

4. The teacher determines which solutions are correct. A group is eliminated when its solution is incorrect.

5. Continue playing rounds, repeating Procedures 1–4. The game is over when only one group remains.

6. Each group that was eliminated chooses one of its players to tell the class what equation they solved incorrectly. Other class members offer suggestions for correctly solving the equation.

Cooperative-Learning Activity
5.5 Magic Triangles

Group members: 2–4

Materials: paper, pencil

Responsibilities: Create a magic triangle.

Preparation: The binomials $2x - 1$, $3x + 1$, $x + 4$, $4x - 2$, $x + 2$, and $3x + 3$ have been arranged on the triangle so that the sum of the binomials on each side is $6x + 4$. Because the sums of the three expressions on each side are all equal, this triangle is an example of a *magic triangle*.

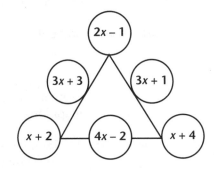

$(x + 2) + (3x + 3) + (2x - 1) = 6x + 4$
$(2x - 1) + (3x + 1) + (x + 4) = 6x + 4$
$(x + 2) + (4x - 2) + (x + 4) = 6x + 4$

Procedures: 1. Each group tries to form other magic triangles, using the same six binomials in the magic triangle above but in different positions on the triangle. Members may work alone or in pairs. Members should note that although the same binomials can be used to make other magic triangles, the sums along the sides of the new triangles will not necessarily be $6x + 4$.

2. When any members believe that they have formed a magic triangle, they should present the triangle to the rest of the group. The other members together verify that the correct binomials have been used and that each side of the triangle has the same sum.

3. The members continue the activity until they have found 5 different magic triangles.

4. All magic triangles found by the group should be recorded.

5. The members should discuss the techniques that they used to find a magic triangle. Why are the sums not all the same? One member of the group should be chosen to write a brief summary of the group's discussion.

_____

_____

_____

_____

_____

Cooperative-Learning Activity
5.6 Graphing Inequalities

Group members: 4

Materials: paper and pencil

Responsibilities: Determine solutions of an inequality, graph the inequality, and write an equivalent inequality.

Preparation: Students will use the numbers in the following list to build inequalities:

$$-4 \quad -3 \quad -2 \quad -1 \quad 0 \quad 1 \quad 2 \quad 3 \quad 4$$

An inequality such as $x < 5$ has the variable term on the left and the constant term on the right. The graph of this inequality is shaded to the left of 5. The graph of $x > 5$ is shaded to the right of 5.

Procedure:

1. Each member chooses a number from the list above to substitute for $\square$ in the statement $\square > x$. Each member writes his or her inequality on a piece of paper.

2. The members exchange papers with a partner.

3. The members then list 5 replacements for x that are solutions to the inequality on the paper they have received.

4. The members graph the 5 numbers they listed in Procedure 3 on a number line.

5. Using the results of step 4, the members graph their inequalities.

6. Each member writes an equivalent inequality for the given inequality. The equivalent inequality should have x on the left side of the inequality symbol and the constant on the right side.

7. The members exchange papers with their partners. The members correct their partner's graphs and equivalent inequalities. The partners should discuss any disagreement they have about the graphs or the inequalities.

8. Repeat Procedures 1–7 for the statement $\square < x$. The members should choose different partners for this exercise.

9. The members summarize the results of the activity and answer the following questions:

    If $a < x$, then x __?__ a. _____

    If $a > x$, then x __?__ a. _____

    Is the graph of $6 < x$ shaded to the left or right of 6? _____

    Is the graph of $6 > x$ shaded to the left or right of 6? _____

    Are the solutions of $6 < x$ and $x < 6$ equivalent? Explain.

    _____

HRW material copyrighted under notice appearing earlier in this work.

Cooperative-Learning Activity
5.7 Solving Inequalities Game

Group members: 3

Materials: Forty-three 3 × 5 index cards: 40 cards numbered with consecutive integers from −20 to 19 inclusive, 1 card with an x written on it, 1 card with a + sign on it, and 1 card with a > symbol on it

Roles: **Leader** shuffles and deals cards to players

2 Contestants play the cards

Preparation: Students will use the index cards to build inequalities such as the following:

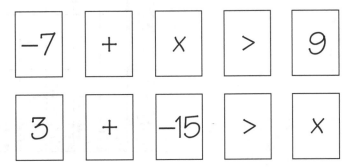

Procedures:
1. The players decide on their roles. The Leader shuffles the 40 numbered cards, places them facedown in one stack, and then turns over two cards from the stack. The Leader arranges the two cards together with the cards having the + sign, the > symbol, and the x variable to form an inequality like either of those shown above.

2. The Leader deals 10 of the cards in the stack, facedown, to each player.

3. Play begins when the Contestants simultaneously turn over the card on the top of their stack. The players decide which cards are solutions to the Leader's inequality.

   - If only one Contestant's card is a solution to the Leader's inequality, then that Contestant takes both cards and places them faceup at the bottom of his or her stack of cards.

   - If both Contestants' cards are solutions to the inequality, then the player with the larger number on his or her card takes both cards. The cards are placed faceup at the bottom of that Contestant's stack of cards.

   - If neither of the Contestants' cards is a solution to the inequality, the Contestants replace their cards at the bottom of the Leader's deck.

4. The Contestants each play 9 more cards this way. The Contestant with the most cards is the winner.

5. The players play the game two more times, rotating the role of Leader so that each member is the Leader once.

6. The players discuss the strategies they used for determining whether their card was a solution to the inequality. How do the strategies for solving $3 + 5 > x$ differ from the strategies for solving $3 + x > 5$?

Cooperative-Learning Activity
6.1 The Distributive Property

Group members: 2

Materials: grid paper, pencil, and pen or a colored pencil

Responsibilities: Use counting and algebraic expressions to find the number of border squares on a square grid.

Preparation: Look at the 7 × 7 grid shown. The border squares are shaded.

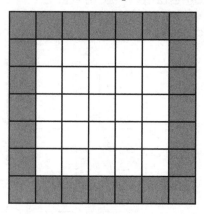

Counting shows that there are 24 shaded border squares. The expressions $4(7 - 2) + 4$ and $(7 \times 4) - 4$ also give the total number of border squares on the grid. Students will use these expressions to find the total number of border squares in other square grids.

Procedures: 1. One member draws a 5 × 5 grid on the grid paper. The other member shades the border squares.

2. The members count the number of border squares. _____

3. The members take turns writing two expressions for the total number of border squares. The members should use the expressions from the 7 × 7 grid given above as a model, substituting 5 for 7 in the expressions.

4. The members discuss why the expressions give the correct number of border squares. On a separate piece of paper, one member writes a short summary of the discussion.

5. The members rotate their roles and repeat Procedures 1–4 for an 8 × 8 grid.

6. The members use the results of the exercises to write two algebraic expressions for finding the number of border squares on a square grid with n sides. Each member is responsible for writing one expression.

7. The members use the Distributive Property to show that these two expressions are equivalent.

8. The members use the expression $4(n - 2) + 4$ to find the total number of border squares in a 3 × 3 grid. Repeat this exercise for a 9 × 9 grid.

Cooperative-Learning Activity
6.2 Modeling Multiplication Equations

Group members: 4

Materials: algebra tiles, calculator, paper, and a pencil

Roles: **Modeler** models an equation with algebra tiles

Solver solves an equation by using algebra tiles

Recorder sketches an algebra tile model of an equation and records the solution to the equation

Checker checks the solution of an equation with a calculator

Preparation: Some multiplication equations can be solved by using algebra tiles. For example, to solve $2x = -6$, model $2x$ with 2 x-tiles and model -6 with 6 negative 1-tiles. Then divide the left side of the equation into two sets, each having one x-tile. Divide the right side into two equal sets of tiles also to find the value of one x-tile.

Thus, $x = -3$

Using the example above as a model, each group will use algebra tiles to solve the following multiplication equations:

$$3a = -9 \qquad 5x = 5 \qquad -4b = 8 \qquad -2n = 10 \qquad -c = -4$$

Procedures: 1. The members choose their roles, which rotate for each equation.

2. The Modeler uses algebra tiles to model an equation.

3. The Solver manipulates the tiles to solve for x and states the solution.

4. The Checker checks the solution by substituting it in the original equation and simplifying the left side of the equation. If the solution is correct, the left side will equal the right side. If the solution is not correct, all members should discuss any discrepancy and decide whether the Solver or the Checker made a mistake.

5. The Recorder draws the algebra-tile model of the equation and its solution.

6. When all of the equations have been solved, the members discuss the procedures that they used to solve the equations. How is solving the equation $2x = -6$ similar to solving the equation $-2x = -6$? How is it different? If $x = -8$, what does $-x$ equal? If $-x = -5$, what does x equal?

_____

_____

Cooperative-Learning Activity
6.3 Factors and Special Products

Group members: 4

Materials: pencil and calculator

Responsibilities: Find and examine special products to determine a pattern for finding the products.

Preparation: Each member of the group should write his or her name in the table below. The members will use the results in the table to write a rule for finding a special product.

Name	Number	Multiply by 7	Multiply by 11	Multiply by 13
1.				
2.				
3.				
4.				

Procedures:

1. Member 1 chooses any three-digit number and records it in the table. The member then uses his or her calculator to find the products indicated and records them in the table.

   • Multiply the number by 7.
   • Multiply this product by 11.
   • Multiply this product by 13.

   At each step, the other members should verify the products on their calculators.

2. Repeat Procedure 1 for all of the members in the group.

3. The group discusses and answers the questions below. All members should participate in the discussion. The members should take turns recording the answers.

   • What is the product of $7 \times 11 \times 13$?
   • Use the results in the table to predict these products:
       527×1001
       111×1001
       989×1001
   • Verify the results on your calculator.
   • Complete this statement. $527 \times 1001 = 527 \times (1000 + \underline{\ ?\ })$
   • Use the results of the completed statement and the Distributive Property to find the product of 527×1001.

4. Each member writes a rule for finding the product of any three-digit number and 1001. The members read their rules to each other and discuss any differences.

5. The members discuss whether switching the order of the factors in Procedure 1 would change the product. Explain.

Cooperative-Learning Activity
6.4 Tug-of-War Proportions

Group members: 4

Materials: paper and pencil

Responsibilities: Use a proportion to solve a problem.

Preparation: Imagine a tug-of-war contest. The participants in this contest are 5 girls, 4 boys, and 1 dog. All of the girls have equal strength. Each girl can pull the tug-of-war rope equally hard. All of the boys have equal strength and can pull equally hard. The 4 boys together can pull the tug-of-war rope as hard as the 5 girls. The dog can pull as hard as 2 girls and 1 boy together.

Procedures: 1. The members choose a partner from their group.

2. Each pair set up a proportion and then solve it to find the number of boys who can tug as hard as 1 girl if 4 boys tug as hard as 5 girls. Both partners in each pair should agree on the solution. _____

3. Let B represent a boy's pulling strength and G represent a girl's pulling strength. The partners use their result from Procedure 1 to discuss and complete this equation: $1G = \underline{\ ?\ } B$.

4. The partners use the result from Procedure 3 to find the number of boys that can pull as hard as 1 dog if 2 girls and 1 boy can tug as hard as 1 dog.

   • Begin with the equation $1D = 2G + 1B$, where D is the dog's pulling strength.
   • Substitute the expression found in Procedure 3 for 1 girl in the right side of the equation.
   • Simplify the right side of the equation. How many boys can pull as hard

   as 1 dog? _____
   The partners should share the responsibilities of this procedure, and both partners should agree on the answers.

5. The partners use the results from Procedures 1–4 to find out which side will win the tug-of-war contest if the dog and 3 girls tug against 4 boys. The partners should discuss how they found the answer. One partner should write a summary of the discussion.

   _____

   _____

   _____

   _____

6. One partner from each pair should read the summary to the other 2 pairs. All members of the group should compare their results and discuss any differences in the procedures they used or the answers they got.

Cooperative-Learning Activity
6.5 Proportions and Sampling

Group members: 5

Materials: 50 small cubes that can be easily marked (such as sugar cubes), marker, paper bag, shoe box, pencil, and paper

Responsibilities: Choose a sample of cubes at random to estimate the total number of cubes.

Preparation: In this activity your group will estimate the total number of cubes in a box by using a sampling technique that involves proportions.

Procedures:
1. One member places the 50 cubes in a shoe box. Another member draws several handfuls of the cubes from the box and places them in a paper bag. The cubes should be drawn randomly from the box and no one should count them.

2. A third member selects a handful of cubes from the bag. A fourth member counts these cubes and marks them with a marker. The fifth member records the number of marked cubes in the first row of the table, and returns the marked cubes to the bag and mixes the cubes well.

3. The members take turns recording their names in the table, selecting a handful of cubes from the bag, and counting the total number of cubes and the number of marked cubes in their hand. Both counts are recorded in the table. Each member returns the cubes he or she has selected to the bag and mixes them up well before the next member makes a selection.

4. Each member finds the totals and the averages for the second and third columns of the table. Round averages to the nearest whole number. One member records the correct totals and averages in the table.

5. Each member uses the results of the activity and the following proportion to estimate the total number of cubes in the box:

$$\frac{\text{number of marked cubes in bag}}{\text{estimate of number of cubes in bag}} = \frac{\text{average number of marked cubes in samples}}{\text{average number of cubes in samples}}$$

6. One member counts the number of cubes in the bag. The members compare their estimate with the actual count. Suppose that the group repeated the experiment again, but this time each member selected five different samples of cubes. Would your estimate be closer to the actual count? Explain.

Number of marked cubes in bag _____		
Student's name	Total number of cubes	Number of marked cubes
1.		
2.		
3.		
4.		
5.		
Total **Average**		

Cooperative-Learning Activity
6.6 Frequency and Percent

Group members: 4

Materials: paper, pencil, and calculators

Roles: **Reader** reads the letters in each word

2 Recorders keep a tally of the frequency of each letter read

Calculator finds the total number of tally marks

Preparation: All legislative powers herein granted shall be vested in a Congress of the United States, which shall consist of a Senate and House of Representatives.

The text above is a copy of Article 1 of the Constitution of the United States. In this activity groups will make a frequency distribution of the occurrence of each letter in Article 1. A frequency distribution shows the number of times each of the letters occurs. To make a frequency distribution, it is often helpful to keep a list or tally of occurrences.

Procedures: 1. The members choose their roles. The Reader reads the letters as they occur in Article 1. As the letters are read, one Recorder keeps a tally of their occurrence in the first half of the article. The second Recorder keeps a tally of the occurrences in the second half of the article. All members ensure that no errors are being made.

2. The Calculator records the total number of tally marks for each letter and sums the tally totals to find the total number of letters in Article 1.

3. Each member uses a calculator to find what percent the total for each letter is of the total number of letters. Round the percents to the nearest tenth of a percent. The Calculator records the correct percents.

Tally	Occurrences	Percent	Tally	Occurrences	Percent	Tally	Occurrences	Percent
A			J			S		
B			K			T		
C			L			U		
D			M			V		
E			N			W		
F			O			X		
G			P			Y		
H			Q			Z		
I			R					
	Total			Total			Total	

4. The members draw a line graph of the frequencies of each letter. The members should take turns graphing the frequencies of the letters. What are the three most frequently used letters in Article 1?

Cooperative-Learning Activity
7.1 Modeling a Puzzle With an Equation

Group members: 4

Materials: calculator, pencil, and paper

Responsibilities: Perform calculations and use algebraic equations to model a number puzzle.

Preparation: Although some people think that number puzzles are magic, they are not. This activity involves writing an equation to model a puzzle that will allow students to determine anyone's birthday.

Procedures:
1. Each group member performs the calculations for each step and records the results in the space provided.

> **Step 1:** Write the number of the month in which you were born. _____
>
> **Step 2:** Multiply by 20. _____
>
> **Step 3:** Add 3. _____
>
> **Step 4:** Multiply by 5. _____
>
> **Step 5:** Add the day of the month in which you were born. _____
>
> **Step 6:** Multiply by 20. _____
>
> **Step 7:** Add 3. _____
>
> **Step 8:** Multiply by 5. _____
>
> **Step 9:** Add the last two digits of the year you were born. _____
>
> **Step 10:** Subtract 1515. _____

2. Each member checks each other's calculations. The members compare their results and discuss any patterns they see. One member is chosen to write a summary of the discussion, which he or she reads to the group for verification.

3. Let m equal the month, d equal the day, and y equal the year. Using these variables, the members take turns writing algebraic expressions for each step of the puzzle. All members should help each other with writing an expression, if necessary. Be sure that everyone agrees on each expression before the next expression is written.

4. The members use the expressions to discuss why the puzzle works. Why is it necessary to subtract 1515? How can you identify the month of a birthday? the year? the day? Is this puzzle magic? Explain. If b equals a person's birthday, what equation models this puzzle? Each member of the group uses the equation to determine the birthday of a member from another group.

Cooperative-Learning Activity
7.2 Modeling a Number Puzzle With an Equation

Group members: 3

Materials: number cube and calculator

Roles: **Roller** rolls a number cube and verifies calculations
Recorder writes down the number rolled and performs computations
Checker uses a calculator to check computations

Preparation: Puzzles in which one person thinks of a number and another person guesses the number are fun because guessing the number appears to involve a trick. This activity use an equation to model the results of this kind of puzzle.

Procedures: 1. The Roller rolls the number cube. The Recorder writes the number, performs the calculations in each step, and records the results. The Checker checks each result. The results are then verified by the Roller.

   Step 1: Roll a number cube. Multiply the result by 2. _____

   Step 2: Add 5. _____

   Step 3: Multiply by 5. _____

   Step 4: Roll a number cube. Add the result. _____

   Step 5: Multiply by 10. _____

   Step 6: Roll a number cube. Add the result. _____

   Step 7: Subtract 250. _____

2. The group discusses and describes the final result. The Recorder records the group's description.

3. The members rotate roles and repeat the activity two more times. Be sure that the Recorder summarizes the group's description of each result.

4. The members choose three different variables to represent the numbers rolled on the first, second, and third roll. Using these variables, the members take turns writing an algebraic expression for each step of the puzzle. All members should help each other when necessary. Be sure all members agree on each expression before the next expression is written.

5. The members use their expressions to explain why this puzzle works. Why is it necessary to subtract 250 in the last step? Let *n* represent the number rolled. Write an equation to model the puzzle.

_____

_____

Cooperative-Learning Activity
7.3 Using Several Equations to Solve a Problem

Group members: 3

Materials: cubes, buttons, algebra tiles, or other small objects, and a pencil

Responsibilities: Model an application involving several equations.

Preparation: Imagine a perfectly balanced scale. The objects on the scale include a pitcher on the left pan and a bottle on the right pan. The bottle is removed from the right pan and replaced by one cup and one saucer. The scale remains perfectly balanced. All objects are removed from the scale. Next, three saucers are placed on the left pan, and two bottles are placed on the right pan. Again, the scale remains perfectly balanced. Can you determine the number of cups needed to balance a pitcher?

Procedures: 1. The members choose four different objects to represent the pitcher, the bottle, the cup, and the saucer. One member records the objects selected as well as the item that each object represents.

_____

2. Using the objects chosen, each member sets up a physical model of one of the conditions below and then draws the model. Then each member writes an equation that represents the condition modeled. Let p equal the weight of one pitcher, b equal the weight of one bottle, c equal the weight of one cup, and s equals the weight of one saucer. The members verify each other's equations.

   a. one pitcher and one bottle balancing the scale _____

   b. one pitcher balancing one cup and one saucer _____

   c. three saucers balancing two bottles _____

3. The members take turns substituting and manipulating their models in order to write a series of equations that will identify the number of cups needed to balance a pitcher.

4. The members review how they solved the problem. One member is chosen to write a summary of the procedures used and to read the summary to the

   class. _____

   _____

   _____

   _____

   _____

HRW material copyrighted under notice appearing earlier in this work.

Cooperative-Learning Activity
7.4 Finding the Greatest Possible Perimeter

Group members: 5

Materials: grid paper, scissors, pencil, and paper

Responsibilities: Draw congruent figures, each with half of the area of a given square.

Preparation: A 4 × 4 grid consists of 16 square units. There are six different ways to divide this square grid into two congruent figures by cutting along the grid lines.

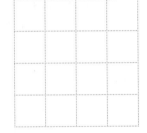

Examine one of the cuts. The shaded figure, which has been cut out of the 4 × 4 grid, has eight 1 × 1 square units, or half of the area of the original grid. Notice that the shaded figure has the same shape as the unshaded figure. The perimeter of the shaded figure is 18.

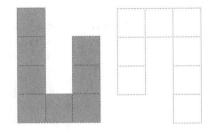

Procedures:

1. Each member tries to draw the four remaining ways that the 4 × 4 grid can be cut into two congruent halves. Use shading to indicate where the grids should be cut. Members may find it helpful to cut out their figures to see whether the shaded and unshaded figures are the same shape. The members compare their drawings, and each member finds the perimeter of his or her shaded figures.

2. The members circle the shapes with the greatest perimeter. What do you think is the greatest possible perimeter of two congruent figures cut from a 4 × 4 grid?

3. Each member tries to draw the six ways that a 6 × 6 grid can be cut into two congruent halves. The members compare their drawings, and each member finds the perimeter of his or her shaded figures.

4. The members discuss the areas and perimeters of the figures they drew in Procedure 3. What is the area of each shaded figure? Which shaded figure has the greatest perimeter? What do you think is the greatest possible perimeter of two congruent figures cut from a 6 × 6 grid?

5. The members discuss the strategies they used for drawing the figures on each of the grids. Did they use the same strategies or different strategies for the two grids? Explain. One member is chosen to take notes of the discussion. The notes should be read to the other members to ensure that the notes are accurate and complete. The member who took the notes writes a summary of the discussion, which is then shared with the rest of the class.

Cooperative-Learning Activity
7.5 The Sides of Triangles

Group members: 3

Materials: scissors, tape, centimeter graph paper, pencil, and paper

Responsibilities: Measure, cut, fold, and tape strips of graph paper together to form triangles.

Preparation: For three line segments to form the sides of a triangle, there must be a specific relationship among their lengths. Can you discover when line segments of given lengths will form a triangle?

To make a triangle with sides of 3 cm, 4 cm, and 5 cm, cut out a 1 cm × 12 cm strip from a sheet of centimeter graph paper. Measure and fold the strip into 3 cm, 4 cm, and 5 cm sections. Then form a triangle by taping the ends of the strip together.

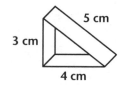

Procedures:

1. Each member selects set A, B, or C. Each member should select a different set of measurements. Note that each set contains three groups of numbers. The members must decide which groups are possible lengths of the sides of a triangle. The members measure, cut, and fold three strips of centimeter graph paper as described in the preparation above, using the lengths in their set. Whenever a triangle can be formed from a strip, the members tape the ends of the strip together.

   Set A: 4 cm, 5 cm, 8 cm; 3 cm, 5 cm, 8 cm; 2 cm 4 cm, 5 cm
   Set B: 6 cm, 6 cm, 6 cm; 6 cm, 7 cm, 6 cm; 3 cm, 4 cm, 8 cm
   Set C: 4 cm, 5 cm, 7 cm; 4 cm, 4 cm, 7 cm; 4 cm, 5 cm, 8 cm

2. Each member describes and records what happened as he or she attempted to form the triangles.

   Set A: _____

   Set B: _____

   Set C: _____

3. Each member lists the measurements for which it was possible to form triangles.

4. Each member compares the sum of the lengths of any two sides to the length of the third side. The members discuss the results. How does the relationship between the lengths relate to the possibility of forming a triangle? One member is selected to record the results of the discussion.

5. Each member writes two sets of numbers on a piece of paper. It should be possible to form a triangle from one set of numbers. It should not be possible to form a triangle from the other set of numbers. The members should exchange papers and verify which set of numbers will form a triangle and which set will not form a triangle.

Cooperative-Learning Activity
7.6 Absolute-Value Slide Rule

Group members: 2

Materials: notebook paper, scissors, and a ruler

Responsibilities: Construct and use an absolute-value slide rule.

Preparation: An equation such as $|x - 2| = 5$ has two solutions, $|5| = 5$ and $|-5| = 5$. Therefore, on the number line x lies 2 units to the right of 5 at 7 or 2 units to the right of -5 at -3. Two scales, one showing integers and the other showing the absolute values of the integers can be used to demonstrate the solutions to this equation.

Procedures:

1. Each partner cuts one 1-inch-by-11-inch strip of notebook paper lengthwise.

2. On the top edge of their strips, the partners place a 0 in the center. One partner writes positive and negative integers along the upper edge of his or her strip, using the ruled lines so the pairs of consecutive integers are equidistant. This strip should be labeled "Number Line." The other partner writes the absolute value of the integers to the left and right of zero, again using the ruled lines as a guide so the numbers are equally spaced. This strip should be labeled "Distance From."

-8	-7	-6	-5	-4	-3	-2	-1	0	1	2	3	4	5	6	7

Number Line

8	7	6	5	4	3	2	1	0	1	2	3	4	5	6	7

Distance From

3. Both partners position their strips as shown, making certain that the scales are aligned.

4. To solve $|x - 2| = 5$, one partner moves the Distance From scale horizontally right until the 0 on the Distance From scale lines up with the 2 on the Number Line scale. The other partner notes where the two 5s on the Distance From scale align with their corresponding values on the Number Line scale. The slide rule demonstrates that the values $x = -3$ and $x = 7$ solve the equation.

5. The partners use their slide rules to find the solutions to the equations below, taking turns manipulating the scales and reading the solutions.

   $|x - 4| = 3$ _____ $|x + 2| = 0$ _____ $|-4 + x| = 2$ _____

6. The partners discuss how they can use their absolute-value slide rule to find the values of x that solve $|x - 1| < 4$. Then the members manipulate the slide rule to find the solutions to the inequality.

7. On a separate piece of paper, the partners discuss the difference between the solutions of absolute-value equations and the solutions of absolute-value inequalities.

Cooperative-Learning Activity
8.1 Exploring Linear Functions

Group members: 4

Materials: paper and pencil

Roles: Calculator calculates sums

Checker checks sums

Recorder records answers

Reporter summarizes group's work

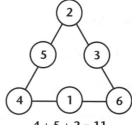

4 + 5 + 2 = 11
2 + 3 + 6 = 11
6 + 1 + 4 = 11

Preparation: The numbers 1 through 6 have been arranged on the vertices and midpoints of the sides of a triangle so that the sum along each side is equal to 11. Note that each number is used exactly once and that the sum of the vertices is 12.

Procedures:

1. The members choose their roles. The members work alone to complete the first triangle. Position the numbers 1, 2, 3, 4, 5, and 6 so that the sum along each side is equal to 9.

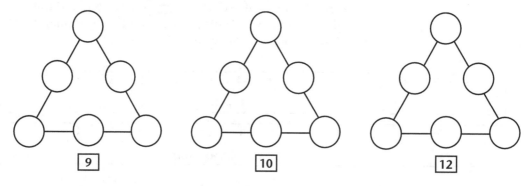

9   10   12

2. When a member believes he or she has completed the triangle, the Calculator finds the sums indicated in the table below. The Checker verifies the sums, and the Recorder records the correct sums in the table.

Sum of the sides (s)	9	10	11	12
Sum of the vertices (v)			12	
Sum of the numbers in the triangle			21	

3. The members repeat Procedures 1 and 2 for the other two triangles. The roles of Calculator, Checker, and Recorder are rotated for each triangle.

4. The members compare the sum of the numbers used in each triangle to the sum along the sides, s, and the sum of the vertices, v, for each triangle. How many sides does a triangle have? How many times is each vertex added to find the sum of all the numbers used in a triangle? The members work together to write a formula that relates the sum of the sides and the sum of the vertices to the total sum. The Reporter verifies that the formula works for each of the triangles.

5. The members discuss whether there is a linear relationship between the sum of the sides and the sum of the vertices. Explain. The Reporter writes a summary of the group's discussion.

Cooperative-Learning Activity
8.2 Slope Cycles

Group members: 3

Materials: ruler, grid paper, plastic sheet or tracing paper, and pencil

Responsibilities: **Leader** oversees group activities and asks teacher for assistance, if necessary

Coordinator makes sure all group members agree on answers

Recorder records answers and summarizes group work

Preparation: The last digit of the first eight multiples of 4 have been plotted on the graph below. The coordinates of the plotted points are (0, 0), (1, 4), (2, 8), (3, 2), (4, 6), (5, 0), (6, 4), and (7, 8). The points have been connected in order on the graph.

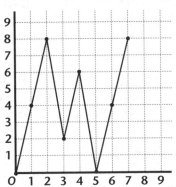

Whole number	Slopes of segments	Length of cycle
3		
5		
6		
7		

In this activity, students will plot the last digit of multiples of other whole numbers and describe the patterns and relationships suggested by the graphs.

Procedure:
1. The members choose their roles. The Leader directs one member to calculate the slope of the line segment between the points (1, 4) and (2, 8). A second member is directed to calculate the slope of the line segment between the points (2, 8) and (3, 2). The members then find the slope of each segment on the graph. The Leader ensures that all members share the task equally. The Coordinator makes sure that all group members agree on the answers, and the Recorder records the slopes.

2. The Leader guides the members as they compare slopes. How are the slopes alike? How are they different? Explain why the graph is not a straight line.

3. The Leader selects one member to trace the section of this graph between the points (0, 0) and (5, 0) onto a sheet of clear plastic or tracing paper. Another member is then selected to slide the tracing along the x-axis. Does the graph appear to repeat in cycles? If so, find the length of the cycle.

4. The members rotate their roles and repeat the procedure for the last digit of the multiples of 3, 5, 6, and 7. First draw a graph for each set of points until the y-values of the first three points have been repeated. Find the slope of each segment on the graph. Make a tracing of the first cycle of the graph, and slide it along the graph to verify that the graph repeats in cycles. State the length of each cycle.

5. The members use the completed table to predict the slope of the next segment on each graph. Each member then calculates the slope to check the prediction.

Cooperative-Learning Activity
8.3 Graphing a Foot Race

Group members: 2–3

Materials: calculator, pencil, and grid paper

Responsibilities: Make a table of points and plot the points on a set of coordinate axes.

Preparation: Alissa, John, and Andy plan to run a race. Alissa is an experienced runner, John rarely runs, and Andy is a moderately good runner. The three friends decided on a way to make the race fair. They decided that Alissa and John will run as a team against Andy. Both Alissa and John will run half of the distance and combine their times, while Andy will run the full distance. The friends agreed on a total distance of 60 meters. Use a graph to determine who will win the race if John runs at an average speed of 1.5 meters per second, Andy runs at an average speed of 3 meters per second, and Alissa runs at an average speed of 6 meters per second.

Procedure: **1.** The members complete the tables by using the equations given in order to find the distance run by each person after t seconds. Each member should complete at least one table. Members should verify each other's work.

Alissa		John and Alissa			Andy	
$6t = d$		$1.5t + 30 = d$			$3t = d$	
time in seconds (t)	distance in meters (d)	John's time in seconds (t)	John and Alissa's combined distance in meters (d)	John and Alissa's combined time in seconds	time in seconds (t)	distance in meters (d)
1		5			0	
2		8			5	
3		10			10	
4		15			15	
5		20			20	

2. The members who completed the tables for Alissa and for John and Alissa, graph the set of points that show John and Alissa's combined time and their combined distance. Graph time on the horizontal axis and graph distance on the vertical axis. The member who completed Andy's table should plot those points on the same coordinate axes. The members should check each other's graphs.

3. The members discuss their tables. Why is 30 added to John's distance? How far has Andy run at the end of 20 seconds? What is the total distance that Alissa and John have run at the end of 20 seconds? Who will win the race? Explain. One member should write a summary of the discussion.

4. The members discuss their graphs. How do Andy's graph and the graph of John and Alissa's combined time show you who would win the race? A member who did not write the summary in Step 3 should write a summary of this discussion.

Cooperative-Learning Activity
8.4 Tic-Tac-Toe

Group members: 2

Materials: grid paper and a pencil

Responsibilities: Play tic-tac-toe on a coordinate grid and write equations to represent winning arrangements.

Preparation: In the game of tic-tac-toe, the first player places an X in any one of nine squares. Then the second player places an O in any of the remaining eight squares. The game continues, with players taking turns, until one player has three marks, either Xs or Os, in a line. The line can be horizontal, vertical, or diagonal. If neither player succeeds in placing three marks in a line, the game ends in a draw. During play, each player tries to block the opponent from placing three marks in a line on the grid.

In this activity, the group will play tic-tac-toe on a coordinate grid, using ordered pairs to designate moves.

Procedures: 1. The following games have been partially played. The members play the remainder of each game, using ordered pairs to designate each move. The members record their moves as they are made.

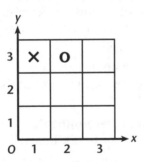 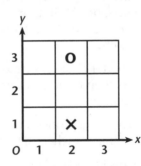 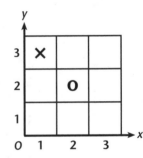

2. The players take turns writing all possible winning arrangements of Xs on a coordinate grid. The location of each X should be labeled with its ordered pair. Hint: One winning arrangement has Xs along a diagonal, with coordinates (1, 3), (2, 2), and (3, 1). There are seven more possible arrangements.

3. The members draw lines through the winning arrangements that they identified in Procedure 2, and then write an equation for each line by using two points on the line. The members check each other's equations.

4. The members find the slope of the diagonal and horizontal lines and verify their answers with each other. Which lines have a slope whose absolute value is 1? What kinds of lines have a slope of zero? Explain. One member should be selected to write a summary of the explanation.

Cooperative-Learning Activity
8.5 Modeling Equations of Lines

Group members: 9

Materials: meter stick or yardstick, chalk or masking tape, and a pencil

Roles: **Plotters** plot a point by standing in the correct position on a model of the coordinate axes

Checkers verify a plotter's position

Writer translates a sentence into an equation

Grapher graphs a linear equation

Recorder records data

Preparation: As the teacher describes a rule for determining a set of points that lie on a line, the members represent some of the points on the line by positioning themselves on a model of the coordinate axes. The members can model the axes by using chalk in a outdoor area or by using masking tape in an indoor area. The axes should be numbered from –5 to 5, and each interval should be about 10 inches.

Procedures:
1. The members choose their roles. The group should have three Plotters, three Checkers, one Writer, one Grapher, and one Recorder. The Plotters and Checkers should form three teams, with one Plotter and one Checker on each team.

2. Each Plotter chooses one x-coordinate between -4 and 4 inclusive. The Plotters then position themselves on the x-axis at the x-coordinate that they have chosen.

3. The teacher describes a rule for determining the y-coordinates of each x-coordinate, such as "Double your x-coordinate. Then subtract three."

4. The Checkers calculate the y-coordinates for their Plotters and announce the ordered pairs for each point. The Plotters move vertically to the position of their ordered pair. The Checkers verify that their Plotters are in the correct position on the axes.

5. While the Checkers and Plotters are determining the positions of the points, the Writer writes an equation for the teacher's rule. Then the Grapher uses the equation to sketch the graph in a table similar to the one below. The Recorder records the rest of the data in the table. The Checkers, Writer, Grapher, and Recorder verify that the sketch of the graph matches the Plotters' model.

Rule description	x	y	Equation	Sketch of graph

6. Repeat Procedures 1–5, rotating roles. The members discuss methods of graphing equations. One member is chosen to write a summary of the methods, which he or she then reads to the class.

HRW material copyrighted under notice appearing earlier in this work.

Cooperative-Learning Activity
8.6 Modeling Lines on a Geoboard

Group members: 3

Materials: geoboard, rubber bands, paper, and a pencil

Roles: **Modeler** models lines on a geoboard

Spokesperson answers the questions for the group

Recorder records the group's answers

Preparation: A geoboard can be used to model linear equations. By stretching two rubber bands between pegs so that the rubber bands are perpendicular to each other, you can model a pair of coordinate axes. Other rubber bands placed on the axes can model equations. The model can be used to explore the slopes of vertical and horizontal lines.

Procedures:
1. The members choose their roles. The Modeler constructs a set of coordinate axes by stretching two rubber bands perpendicular to each other in the middle of the geoboard.

2. The Modeler stretches a rubber band between $(-2, -1)$ and $(2, -1)$. The Spokesperson answers these questions: What is the rise between these two points? What is the run? Tell why the slope for this line is 0. The other members verify the answers, and the Recorder records the correct answers on a piece of paper.

3. The members rotate roles and repeat Procedure 2 for the points $(-2, 2)$ and $(2, 2)$.

4. The members discuss why the slope for a horizontal line is 0. The Recorder records the discussion.

5. The members rotate roles so that all members have roles that they have not had before. The Modeler stretches a rubber band between $(-2, 1)$ and $(-2, -2)$. The Spokesperson answers these questions: What is the rise between these two points? What is the run? Tell why the slope for this line is unusual. The other members verify the answers, and the Recorder records the correct answers.

6. The members take turns modeling a vertical line on the geoboard. Each member finds the slope of his or her line, and the other members verify the slope.

7. The members discuss why the slope for a vertical line is undefined. The Recorder from Procedure 5 records the discussion.

8. The members rotate roles again. The Modeler stretches a rubber band between $(-1, -2)$, and $(1, 2)$. All members find the slope of the line. The Spokesperson identifies the y-intercept of the equation. The Modeler stretches a rubber band between $(-2, -1)$ and $(2, 1)$. Each member finds the slope of the line. The Spokesperson determines the point that both lines have in common. The members together write the slope-intercept equation for each line. The Recorder records all of the answers for this exercise.

Cooperative-Learning Activity
8.7 Optical Illusions

Group members: 3

Materials: ruler, grid paper, and a clear plastic sheet or tracing paper

Roles: **Leader** oversees group activities and asks the teacher for assistance, if necessary

Coordinator makes sure all group members agree on answers

Recorder records answers and summarizes the group's work

Preparation: Parallel lines can be used to create false pictures, or optical illusions. In an optical illusion, the information seen by the eye is misleading, so what the eye sees may not be what is actually represented. Use the following optical illusions for this activity:

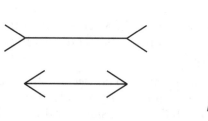

Figure 1

Figure 2

Figure 3

Procedures: **1.** The members choose their roles. The members examine the horizontal line segments in Figure 1 and decide which line segment looks shorter. The members measure the segments. Was your guess correct?

_____

Hint: The arrows are confusing. These symbols may cause the mind to interpret the visual information incorrectly. Be sure to treat visual representations cautiously. It is always safer to rely on correct reasoning rather than what you think you see.

2. The members rotate their roles, examine Figure 2, and answer the following questions:

• Line x intersects two vertical lines. Which line, p or q, is the continuation of line x? (To verify your answer, trace Figure 2 and place it on coordinate axes. Then find the slope of lines x, p, and q.)

• Was your guess correct?

3. The members rotate their roles again, examine Figure 3, and answer the following questions:

• Look at the horizontal lines. Is one line longer than the other? (Measure the lines to verify your answer.)

• Trace Figure 3 and place it on coordinate axes. Find the slope of each non-horizontal line. Discuss how the figure was constructed so that its shape fools the eye.

Cooperative-Learning Activity
8.8 Modeling Linear Inequalities

Group members: 2–4

Materials: small box and four different kinds of small objects, such as cubes, buttons, or algebra tiles

Responsibilities: Write variable expressions to represent unknown quantities and solve an inequality.

Preparation: A box of assorted chocolates has 19 filled chocolates. The fillings include raspberry, hazelnut, almond, and coconut. There are twice as many raspberry as hazelnut-filled chocolates. There are two fewer almond than hazelnut. And there are three fewer coconut than raspberry. Find the number of chocolates with each kind of filling. Then decide how many chocolates you have to take from the box to have at least two with the same kind of filling.

Procedures:

1. One member chooses a variable to represent the number of hazelnut-filled chocolates and records it in the table. The other members take turns filling in the variable expressions for the other fillings. The members verify each other's expressions.

Type of filling	Variable expression	Number
hazelnut		
raspberry		
almond		
coconut		

2. Each member writes an equation to find the number of hazelnut-filled chocolates and then solves the equation.

3. Each member uses the variable expression that he or she entered in the table to find the number of chocolates represented by that expression. The members record their numbers in the table.

4. Each member chooses a different object to represent his or her type of chocolate. The members model the number of each kind of chocolate with the objects they have chosen.

5. One member places the objects in a small box and shakes the box to mix up the objects. Another member is selected as the Recorder. The members take turns selecting objects from the box without looking at them. When the objects are selected, they are given to the Recorder, who records them and replaces them in the box, except for the first object, which the Recorder keeps. When two of the same objects have been selected, the Recorder announces this to the members and replaces all of the objects in the box. Repeat this activity several more times.

6. The members examine their results and discuss the maximum number of selections they have to make from the box to get at least two of the same object. Write an inequality to express this situation.

Cooperative-Learning Activity
9.1 A Diagram for Systems of Equations

Group members: 3

Materials: straightedge, pencil, and paper

Roles: **Leader** oversees group activities and asks teacher for assistance, if necessary

Recorder records answers and summarizes group work

Coordinator makes sure all group members agree on answers

Preparation: An express mail company transports mail from New York City to Montreal. Travel time between the two cities is 9 hours. The company's express vans leave daily from both cities every hour on the hour, starting at 5:00 A.M. and continuing until 8:00 P.M. If a van leaves at 5:00 A.M. from New York City, how many vans will it pass on its way to Montreal?

This real-world problem, which involves a system of equations, can be modeled with a diagram showing the departure and arrival time of each van. A partial diagram is shown.

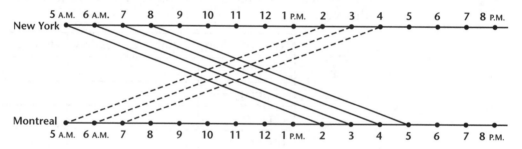

Notice that the first solid line represents the first van leaving New York City at 5:00 A.M. and arriving in Montreal at 2:00 P.M. The first dashed line represents the first van leaving Montreal at 5:00 A.M. and arriving in New York City at 2:00 P.M.

Procedures: 1. Each member chooses a role. With the Leader directing the activity, the members take turns completing the diagram. The Coordinator makes sure that the members all agree that the diagram has been completed correctly.

2. The Coordinator and the Reporter describe what they notice about the slopes of the solid lines in the diagram. The Leader describes what he or she notices about the slopes of the dashed lines in the diagram. Explain why this makes sense. The Coordinator ensures that all members agree with the explanation, and the Recorder records the explanation.

3. The Leader guides the members in a discussion of what the point of intersection of each pair of lines represents. When the Coordinator is sure that all members agree on the answer, the Recorder records the group's answer.

4. Each member determines how many vans the first van headed to Montreal will pass along the way. The Recorder records the group's answer.

5. Each member chooses a van and uses the diagram to demonstrate how many vans the chosen van will pass along its route.

Cooperative-Learning Activity
9.2 Role Playing to Solve a System of Equations

Group members: 4

Materials: "play" dollar bill, paper, and a pencil

Roles: Players   play the role of a person listed below and solve a system of equations

Preparation: Jake owes Oko one dollar. Oko owes Luc one dollar, and Luc owes Jake one dollar. Jake decided to settle his debt. He asked Emily for a loan of one dollar, which Emily gave to him. Jake used the dollar to pay Oko the dollar he owed her. Then Oko paid Luc the dollar, and Luc paid Jake the dollar. Jake then paid Emily the dollar. Have all the debts been paid?

This problem can be solved by role playing and by using a system of linear equations with four variables.

Procedures:
1. Each player assumes the role of one of the people in the problem. The Players describe the situation by using a model of a dollar bill to pay off their debts. Beginning and ending with Jake, as the Players pay a debt, they record the amount they still owe to each of the other Players.

Owes	Jake	Oko	Luc	Emily
Jake				
Oko				
Luc				
Emily				

2. The Players discuss the final result of the transactions. How does the table confirm the results? The Player representing Emily records the group's answers.

3. The Players check the result by solving a system of linear equations. Because Jake owes Oko one dollar, if j represents Jake, and o represents Oko, the equation $j = -o$ can be used to model this relationship. Using the variables j, o, l, and e, each Player writes a similar equation for each person to whom he or she owes a dollar.

4. The Player representing Emily verifies that the equations are correct and then records them.

5. The Players work in teams of two, using substitution to solve the system of equations written in Procedure 4. The teams compare their solutions. Do the solutions of the equations agree with the results of the role playing?

6. One member of each team of Players writes an explanation of the methods they used to solve the system of equations. The other members proofread the explanation and make any necessary changes.

7. The teams read each other's summaries and discuss similarities and differences in the methods they used.

Cooperative-Learning Activity
9.3 Modeling Systems of Equations

Group members :   4

Materials: positive and negative *x*-tiles, positive and negative *y*-tiles, pencil, and paper

Roles: **Writer**   writes a system of equations

Modeler   uses algebra tiles to model and solve a system of equations

Coordinator   makes sure all members agree on answers

Preparation: On a 20-question test, one point is deducted for each incorrect answer and three points are credited for each correct answer. If all of the questions are answered and the score on the test is zero, how many answers were correct?

Since this problem can be modeled by a system of linear equations, both algebra tiles and the elimination method of solving a system of linear equations can be used to find the number of correct answers.

Procedures: 1.   The members choose their roles. The group should have one Writer, two Modelers, and one Coordinator.

2.   Let *x* equal the number of correct answers, and let *y* equal the number of incorrect answers. The Writer writes a system of equations that describes the total number of questions on the test and the number of correct and incorrect answers if the score is zero. The Coordinator ensures that the other members agree with the system of equations.

_____

3.   The Modelers model this system of equations by using the *x*-tiles and the *y*-tiles. One Modeler models one equation, and the other Modeler models the other equation. The Modelers combine the two models, and the Coordinator verifies that all members agree with the model.

4.   The members decide which neutral pairs to remove and one Modeler removes them. Then the Writer writes the new equation. The Coordinator makes sure that all members agree with the new model and the equation.

_____

5.   The Writer solves the resulting equation and uses substitution to find the value of the variable that was eliminated by removing neutral pairs. The other members guide the Writer.

_____

6.   Each member substitutes the values of *x* and *y* into each of the original equations to confirm that the solution is correct. The Coordinator records

the number of correct answers. _____

7.   Each member uses the elimination method to solve the system in Procedure 2. The members check their solutions with each other.

HRW material copyrighted under notice appearing earlier in this work.

Cooperative-Learning Activity
9.4 Elimination by Multiplication

Group members: 2

Materials: paper, pencil, and a calculator (optional)

Roles: **Solver** solves systems of equations and examines the equations and the solutions for a pattern

Recorder records the group's answers and discussions

Preparation: These systems of equations with two unknowns have an interesting property. If you can discover this property, you can describe other systems of linear equations that share this property.

$$\begin{cases} x + 2y = 3 \\ 7x + 9y = 11 \end{cases} \qquad \begin{cases} -x + y = 3 \\ 5x + 7y = 9 \end{cases} \qquad \begin{cases} -2x - y = 0 \\ x + 2y = 3 \end{cases}$$

$$\begin{cases} 3x - y = -5 \\ 12x + 15y = 18 \end{cases} \qquad \begin{cases} 6x + 9y = 12 \\ 15x + 18y = 21 \end{cases} \qquad \begin{cases} 2x - y = -4 \\ 7x + 10y = 13 \end{cases}$$

Procedures:

1. Each member is both a Solver and a Recorder.

2. Each Solver solves three systems of equations.

3. The Solvers compare their solutions and describe what they notice about the solutions. Recorder 1 records the discussion.

   _____

   _____

   _____

4. The members examine the coefficient of the *x*-term, the coefficient of the *y*-term, and the constant term for each equation. How are these numbers related? Is this relationship true for all twelve equations? Recorder 2 records the answers.

   _____

   _____

   _____

5. The Solvers each write a system of linear equations with two unknowns that have the property identified in Procedure 4. The Solvers then check their partner's solution by solving the system of equations.

   _____

   _____

   _____

Cooperative-Learning Activity
9.5 Systems of Linear Inequalities

Group members: 2–3

Materials: paper and a pencil

Responsibilities: Determine a system of linear inequalities when given the graph of the solution set.

Preparation: Each of the geometric figures below is the graph of the solution set of a system of linear inequalities. The inequalities in each system determine the boundaries of the figure, so by identifying the boundary lines of the graph, you can identify the system of inequalities.

A
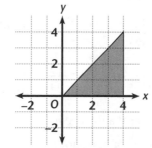

B
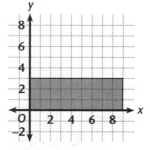

C
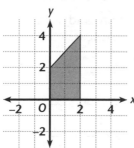

D
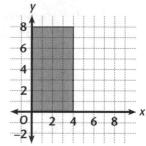

Procedures:

1. The members examine graph A, and select one of the boundaries of the figure. Each member identifies two points on this boundary and then uses the formula $y - y_1 = m(x - x_1)$ to write the equation of the boundary line. The members verify each other's equations. The equation of every boundary line for this figure should be written in a similar manner. The members then discuss the system of inequalities represented by the graph, and one member is chosen to record the system.

2. The members repeat Procedure 1 for the remaining graphs. The members take turns recording the system of inequalities.

3. The members discuss how the solution set of a system of inequalities is graphed. How do you determine whether to shade above or below a boundary line? How is the graph of a system of linear inequalities different from the graph of a system of linear equations? One member is selected to write a summary of the discussion. The other members read the summary and edit it, if necessary.

Cooperative-Learning Activity
10.1 Previewing Exponential Decay Functions

Group Members: 3

Materials: 30 small cubes, paper cup, paper, and grid paper

Roles: **Recorder** records the results, removes the *D* cubes, and totals the data

Plotter plots the data points and draws the curve

Checker verifies all group input and writes description of graph

Preparation: Each team will simulate an exponential decay experiment.

Procedures: **1.** Each member chooses a role. One member distributes the small cubes so that each member has 10 cubes. Members should mark one side of each of the 10 cubes with a letter *D*. The *D* stands for decay. Place the cubes in the paper cup. Shake and roll the cubes.

2. The Recorder counts the number of cubes showing the letter *D* faceup and the number of remaining cubes, and records those numbers in a table similar to the one below. The Recorder then removes the cubes showing the letter *D*.

Number of rolls (*x*)	Number of *D* cubes	Number of Remaining cubes (*r*)

3. Roll the remaining cubes, counting and removing the cubes showing the letter *D*. Continue this procedure until 6 or fewer cubes remain.

4. The Plotter plots the data points (*x*, *r*), where *x* is the number of the roll and *r* is the number of cubes remaining. The Plotter then draws a curve so that it passes close to most of the data points.

5. After discussing the findings of the experiment with the group, the Writer describes the graph.

_____

_____

_____

Cooperative-Learning Activity
10.2 Constructing Parabolas

Group Members: 2

Materials: paper, colored pencils, ruler, and a protractor

Roles: **Connector** connects points that form a parabola

Shader shades in the parabola

Preparation: Each team will construct a parabola. In the figure shown, an angle has been constructed and 11 equally spaced points are located and marked on each side.

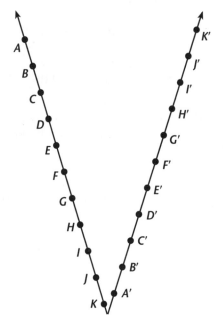

Procedures: **1.** The Connector will connect point *A* to point *A'*, *B* to *B'*, and *C* to *C'*. Continue the procedure until all corresponding points are connected. The Shader then uses a colored pencil to highlight the curve formed by the connected segments.

2. Many different variations of this basic construction are made possible by changing the angle between the line segments or the distance between the points and by combining angles. On a separate piece of paper, each member constructs a different angle and marks equally spaced points on each side of the angle. Then try to find a parabola within the angle you constructed.

3. As a group discuss your findings. What are some basic characteristics of parabolas? Explain.

_____

_____

_____

Cooperative-Learning Activity
10.3 Previewing Reciprocal Functions

Group Members: 2

Materials: colored pencils, paper, and a ruler

Responsibilities: Sketch the graph of the reciprocal of the function.

Preparation: In order to find the reciprocal of a function, first identify at least 4 coordinates on the graph. Create a table that identifies the x-values and the y-values. Then create another column that represents the reciprocal of y, $\frac{1}{y}$. In order to graph the reciprocal of the function, plot the values $\left(x, \frac{1}{y}\right)$.

Procedures: 1. Each team will graph the reciprocal of each function by plotting the new ordered pairs on the graph and then connecting those points.

2. Use the graph of each function to create a table of the ordered pairs. Then create a table of the reciprocal's ordered pairs and sketch the graph of the reciprocal on the same graph as the original function. Identify at least 4 ordered pairs for each graph below:

a. $y = x + 2, x \geq 1$

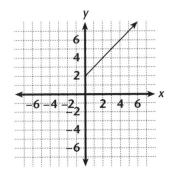

b. $y = x^2, x \geq 1$

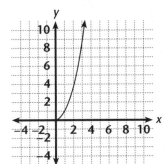

c. $y = 3$

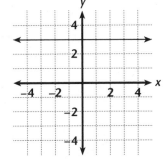

3. As a group, compare the function and the graph of its reciprocal. Describe what you notice.

_____

_____

_____

_____

Cooperative-Learning Activity
10.4 Planting as a Function of Temperature

Group Members: 3

Materials: grid paper, pencils, and a ruler

Roles: **Recorder** creates a table and records the groups findings

Grapher graphs the temperature as a function of the zone number

Verifier verifies all group input and output

Preparation: Winter temperature is the critical factor used to indicate which fruit and nut varieties will grow successfully in a region. Planting dates for some vegetables and many herbs hinge on the last spring's frost or the first fall's frost. The U.S. Department of Agriculture divides the map of the United States into regions called hardiness zones. The map is based on average annual minimum temperatures.

Procedures: 1. As a group, examine the map displaying the hardiness zones. Look for patterns that can be represented by functions.

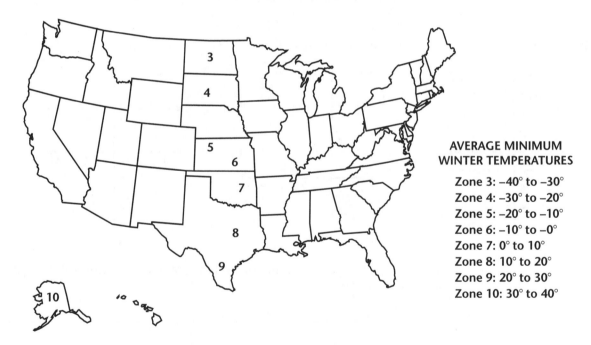

AVERAGE MINIMUM WINTER TEMPERATURES
Zone 3: –40° to –30°
Zone 4: –30° to –20°
Zone 5: –20° to –10°
Zone 6: –10° to –0°
Zone 7: 0° to 10°
Zone 8: 10° to 20°
Zone 9: 20° to 30°
Zone 10: 30° to 40°

2. With assistance from the group, the Writer creates a table to show the zone number and the average minimum winter temperatures.

3. The Grapher creates a graph of the temperature as a function of the zone number.

4. As a group, discuss and describe the graph of your function.

_____

_____

Cooperative-Learning Activity
10.5 Identifying Types of Functions With Patterns

Group Members: 2

Materials: 4 black counters and 4 white counters

Responsibilities: Four black counters are placed in squares 2–5, and 4 white counters are placed in squares 7–10. The remaining squares are empty. The goal of the game is to switch the colored counters from one side to the other. Only two moves are allowed. You may slide one counter to a space or jump over one counter to a space. The group should create a table to record the patterns that exist within the game. The table should have two columns, one representing the number of each color and the other column representing the pattern of moves. This table will be used starting with Procedure 2.

Empty					Empty					Empty
1	2	3	4	5	6	7	8	9	10	11

Procedures:

1. The game begins with the first player moving the counters one at a time, while the second player records the total number of moves needed by the first player to change the counters over. The second player then takes his or her turn to play. The first player records the total number of moves needed by the second player. The player that makes the least number of moves wins the round. In order to see the pattern that determines the least number of moves needed, you will need to play the game with increasing number of counters.

2. Place one counter in the squares marked 5 and 7, and record the color of each counter as you move them one at a time. Repeat the game. Describe the pattern that you notice.

   _____

   _____

3. Begin with 2 counters on each side, and then 3 counters on each side. Record the color of each counter as you move them one at a time.

   What is the least number of moves needed? _____

   _____

4. Write a rule for calculating the least number of moves. Then replay the game with 4 counters on each side to check your rule.

   _____

   _____

Cooperative-Learning Activity
10.6 Exploring Transformations With Different Shapes

Group Members: 2

Materials: mirrors and paper

Figure 1

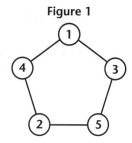

Responsibilities: Find the total number of possible rotations and reflections.

Procedures: Five circles are arranged in a pentagon. How can you place the integers 1, 2, 3, 4, and 5 in the circles so that no two consecutive integers occupy circles connected together by a line? One solution is shown in Figure 1.

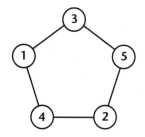 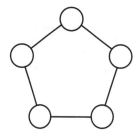 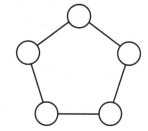 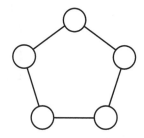

1. If the circle design in Figure 1 is rotated 72°, how many more solutions can be found? One solution is done for you. Take turns recording your solutions.

2. If a mirror is placed on the line of symmetry, the reflections are also solutions. Figure 2 is a reflection of Figure 1. As a group, complete the remaining solutions by similar reflections.

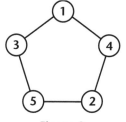

Figure 2

For this design, how many solutions are there? _____

3. As a group, find as many solutions as possible for each design below. Use rotations and reflections to check your new solutions. Draw the outcomes on a piece of paper.

a.

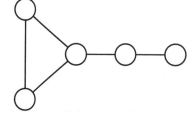
_____

b.
○———○———○———○———○ _____

Cooperative-Learning Activity
11.1 Animal Bar Graphs

Group Members: 3–4

Materials: 8.5-oz box of animal crackers for each group member, different colored pencils, grid paper, and a ruler

Responsibilities: Use animal crackers to create a bar graph.

Preparation: As a group, you will create a bar graph to represent the frequency of different animals in each member's box of animal crackers. A bar graph is a graph in which information is shown by means of rectangular bars.

Procedures: 1. Each member opens a box of animal crackers that contains a variety of animal shapes. Each member should separate the crackers in their box by shape and count how many there are of each shape. All broken crackers should be counted as a broken shape.

2. Tally the results from each box in a table created by the group. The table should have two columns, one representing animal shape and one representing number.

3. Now display the results from your table on a combined bar graph as follows:

- On the horizontal axis, evenly space the name of each animal shape contained in the animal cracker box, such as lion, bear, or cow. Leave enough room for the result from each member's count to be displayed side by side.
- On the vertical scale, evenly space all whole numbers from 0 to the number representing the highest count.
- Draw a bar graph of the count for each animal shape with a colored pencil. Each member will use a different color from the others in the group.

4. As a group, compare the results for each animal shape and answer the following questions:

- Are the counts for the various forms the same or approximately the same

  for all the boxes? _____

- Were there a similar number of broken cookies in each box?_____

- Do you think the results would be the same if you added more boxes?_____

  _____

- Write a summary of the group's findings._____

  _____

  _____

  _____

Cooperative Learning
11.2 Measures of Central Tendency Game

Group Members: 2

Materials: game board, penny, 32 square tiles, and calculators

Responsibilities: Prepare and play a game.

Preparation: Imagine that a ball is rolled through the triangular array of pegs shown in the diagram below. The path that the ball takes as it moves through the array will be determined by the direction change caused when it comes in contact with any of the pegs. The ball has an equal chance to go either to the left or to the right with each contact. As a result, it can end at any finish bin—A, B, C, D, E, or F. One possible path that a ball might travel is shown in the diagram.

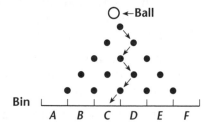

To model this game, a tile will represent the ball. A coin toss will substitute for a peg contact to determine which direction the ball will go. If the coin shows heads, the ball goes right. If the coin shows tails, the ball goes left.

Procedures: 1. Make a game board representing the triangular array of pegs. Use squares in place of the pegs.

2. Before playing the game, make a prediction of how many times a tile will end up at each finish bin, A, B, C, D, E, and F, based on 32 tiles going through the array. Record your predictions in a table created by the group, representing the bins, A–F, in one column and the total number of finishes in the other. Then find the mean, median, mode, and range of the data you predicted.

3. Begin with the first player tossing the coin, while the second player moves the tile. Sixteen tiles are played. The process is repeated, but this time the second player tosses the coin while the first player moves the tiles.

4. After each trial, place a tally mark representing the finish bin in the table. After the 32nd trial, record the total for each bin in a separate table created for the actual results. Then calculate the mean, median, mode, and range of the actual data, and compare the experimental result with your predictions. Describe what you notice.

_____

_____

_____

Cooperative-Learning Activity
11.3 Graphing Data With Stem-and-Leaf Plots

Group Members: 6–8

Materials: 1 box of small-sized elbow macaroni, a medium to large container, and paper

Responsibilities: Create a stem-and-leaf plot.

Preparation: Each group member will see how closely they can estimate the number of elbow macaroni in a handful.

Procedures:

1. Empty the box of elbow macaroni into the container.

2. Each group member estimates how many pieces are in a handful. The estimated number is recorded in the stem-and-leaf plot, arranging the data from least to greatest.

Estimate of pieces in a handful		Average count of pieces in a handful	
Stems	Leaves	Stems	Leaves

3. Each group member scoops out a handful of macaroni and counts the number of pieces. These pieces are returned to the container. This process is repeated two times. An average is calculated for the three handfuls.

4. The average count is recorded in the stem-and-leaf plot above, which orders the data from least to greatest. Then compare the range, median, and mean of the data for the estimate and for the actual counts.

5. On a separate piece of paper, answer the following questions: Did your group overestimate or underestimate the actual number of pieces in a handful? If you repeated the experiment again, would your estimate be more accurate?

Cooperative-Learning Activity
11.4 Circle Graphs

Group Members: 2

Materials: a pencil

Responsibilities: Use circle graphs to establish patterns.

Preparation: A garden club decided to create a circular garden with different colors of a flower called impatiens. They want to plant 8 different colors of impatiens in equally sized plots of land. The diagram shows how the colors will be arranged.

The garden club members claim that the garden will be not only very beautiful, but also magical. Are they right in their thinking?

Procedures: Choose your favorite color from the 8 that will be in the garden. Then spell out that color while your partner taps the diagram for each letter you spell. Your partner should always begin to tap on the color cranberry and then follow the line segment upward to the color silverdust, where your partner taps on the second letter you spell. Continue along the line segments, following the arrows to each of the other colors, for each additional letter in the color you chose. Switch roles for your partner's favorite color. Then answer the following questions:

1. When you have spelled all the letters in a color, where does the tapping finish?

_____

2. Repeat this pattern for all the colors of impatiens. Describe the results.

_____

_____

HRW material copyrighted under notice appearing earlier in this work.

Cooperative-Learning Activity
11.5 Scatter Plots and Correlation

Group Members: 2

Materials: 20-cm ruler showing millimeters, assortment of 12 different small objects (paper clip, pencil, eraser, coins, keys, etc.), 2 different colored pencils, and grid paper

Responsibilities: Use scatter plots to determine correlation.

Preparation: As a group, you will estimate and then measure small objects. You will compare your estimates and measurements by making a scatter plot of the estimates and the actual measurements of the objects. A scatter plot is a graph that shows the relationship between quantities.

Procedures:

1. Place the small objects on a flat surface.

2. Each group member estimates the length of a selected object in millimeters. The median estimate is recorded in a chart created by the group that has three columns, one representing the object, one representing the median estimate, and one representing the actual measurement. A ruler is used to measure the object, and the actual measurement is entered in the chart.

3. This process of estimating and measuring is repeated until all of the lengths of the objects have been recorded in the chart.

4. Create a graph with the horizontal axis marked Measurement and the vertical axis marked Estimate. Form ordered pairs from your data in the chart, and display the information in a scatter plot.

5. As a group answer the following questions:

   a. Did your group consistently overestimate or underestimate? Explain how you interpret your graph according to your group's estimation skills.

   _____

   _____

   b. Did your group's estimation skills improve as the activity progressed? Explain.

   _____

   _____

Cooperative-Learning Activity
11.6 Finding Lines of Best Fit

Group Members: 4–6

Materials: one 8-oz can of pear halves for each student, scale, plastic container, strainer, paper towels, can opener, and grid paper

Responsibilities: Create a scatter plot.

Preparation: As a group, you will approximate the quantity of pears contained in an 8-oz can of pear halves.

Procedures:

1. Weigh a container that will hold the pear halves and syrup. Open a can of pear halves, and pour the pear halves and syrup into the container.

2. Weigh the container with the pear halves and syrup. Then calculate the weight of the pear halves and syrup by subtracting the weight of the container from the weight of the container, pear halves, and syrup. Record the results in a chart that contains three columns, one representing each group member's name, one representing the weight of pears, and one representing the weight of pears with syrup.

3. Drain off the syrup, and then place the pear halves in the container. Calculate the weight of the pear halves by subtracting the weight of the container. Record the weight of the pear halves in the chart. What percent of the total weight (pears and syrup) is the weight of the pears?

   _____

4. Record the weight of the pears and syrup and record all of the data from your group in the chart.

5. Form ordered pairs from the data in the chart, and display the information in a scatter plot constructed by the group on a piece of grid paper. Then, as a group, answer the following questions:

   Do the points appear to lie on a straight line? _____

   Is the weight of the pears approximately the same for each can? _____

   _____

   Use what you have learned from this activity to write a generalization about what consumers should know when they are purchasing canned food.

   _____

   _____

Cooperative-Learning Activity
12.1 Exploring Circles With Cans

Group Members: 3

Materials: centimeter grid paper, centimeter ruler, compass, calculator, and small, medium, and large cans for each group member

Responsibilities: Make measurements and estimates.

Preparation: Each member of the group measures the radius of a circle and then estimates the area. The data is graphed to study the relationship between the radius of a circle and its area.

Procedures: 1. Use the centimeter ruler to find the diameter of each of the three circular cans. Measure the diameter to a tenth of a centimeter, divide the diameter by 2, and round to two significant digits to find the radius. Each member will record their result in the chart below.

Size	Radius	Area
Small		
Medium		
Large		

2. Draw an outline of the base of the can on centimeter grid paper. Estimate the area by counting squares and piecing together parts of squares to estimate whole squares. Repeat this process for the other two cans. Record the results in the chart. After each member has completed this process, examine each member's set of measurements as a group. Discuss why they are different.

_____

_____

3. Form ordered pairs from the group data in the chart and display the information as a scatter plot on grid paper. When constructing the scatter plot, use the *x*-axis for the "Radius," and the *y*-axis for the "Area." As a group, examine the scatter plot. Draw the curve that you think best fits the data and describe the curve.

_____

_____

4. Explain how you can use your curve of best fit to estimate the area of a circle if you know the radius.

_____

_____

Cooperative-Learning Activity
12.2 Exploring Surface Area and Volume

Group Members: 2–3

Materials: cube container (1-liter capacity) and 1 liter of colored sand

Responsibilities: As a group, you will tilt a cube container with sand to model the shapes of its cross sections.

Preparation: A cross section of a solid is the plane figure formed by the intersection of the solid and a plane.

Procedures:

1. Fill the cube container with sand approximately 4 centimeters deep.

2. Place the container flat on the table. Notice the polygon cross section formed by the top surface of the sand. The polygon is a square.

3. Tilt the container at a 45° angle to the table. Look at the top surface of the sand. As a group, describe the surface.

   _____

   _____

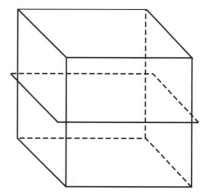

4. Describe how to position the container so that the polygon cross section is a trapezoid.

   _____

   _____

5. Describe how to position the container so that the top surface of the sand has the following shapes, if possible:

   a. a rectangle that is not a square _____

   b. a hexagon _____

   c. an octagon _____

   d. a parallelogram _____

   e. an equilateral triangle _____

   f. a right triangle _____

Cooperative-Learning Activity
12.3 Prisms and Their Nets

Group Members: 2–3

Materials: small cubes, small removable adhesive labels, and paper

Responsibilities: Make estimates and compare prism nets.

Preparation: In this activity, each group member will estimate the result of using certain combinations of color on cube faces. Then each member will color the cube faces in those combinations. Each member's estimates will be compared with the actual result.

Procedures: 1. Each member will visualize a cube and estimate the smallest number of colors needed to paint it so that no two adjacent faces are the same color. Record each member's estimate.

_____

2. Each member will then choose a cube and colored labels. Using the smallest number of colors, place the labels on the cube so that no two adjacent faces have the same colored label. Draw a net of your solution on a piece of paper.

3. Compare your estimate with your result by using cubes and colored labels.

4. Compare your results with each member of your group. Agree on a solution and draw that net in the space provided below.

5. Suppose that you could paint the cube faces with four different colors so that no two adjacent faces have the same color. Visualize the number of differently colored cubes that could be created. Record each member's estimate.

_____

6. Choose four different colored labels. Create as many different cubes as you can. Draw the nets on a separate piece of paper.

7. Compare your solution with those of the other group members. How many distinct ways are there of painting a cube with four different colors so that no two adjacent faces have the same color?

_____

Cooperative-Learning Activity
12.4 Cylinders

Group Members: 3

Materials: 3 small cylindrical objects of different sizes for each team member, centimeter grid paper, tape, a centimeter ruler, and scissors

Responsibilities: Make estimates of the surface area for different cylindrical objects.

Preparation: As a group, follow the procedures below to compare your estimates of the surface area of a cylinder with an actual measurement.

Procedures:
1. Each group member chooses three different-sized cylinders.

2. Outline the circular base of your first cylinder on centimeter grid paper twice, and then cut out both circles. Cover the rectangular lateral surface of the cylinder with centimeter grid paper. Be sure that the entire lateral surface is covered. Mark the grid paper at the point of overlap. Trim off any excess paper.

3. Lay the three pieces of paper on a flat surface. Estimate the surface area by counting squares and piecing together parts of squares to estimate whole squares. Record the result in the chart below.

Cylindrical Object	Estimated Surface Area	Measured Surface Area

4. Use a centimeter ruler to find the diameter and height of each cylinder. Then use these measurements and the formula, $SA = L + 2B$ or $SA = 2\pi rh + 2\pi r^2$, to find the surface area of each cylinder. Record the results in the chart.

5. Record each member's set of measurements and calculations in the chart.

6. As a group, form ordered pairs from the data in the chart and display the information as a scatter plot.

7. Examine the scatter plot. Do the points appear to lie on a straight line? Explain how you can interpret the scatter plot in terms of your group's measurement skills.

_____

_____

_____

Cooperative-Learning Activity
12.5 Volumes of Cones and Pyramids

Group Members: 4–6

Materials: paper, compass, protractor, scissors, centimeter ruler, tape, measuring cup, medium-sized container, and rice

Responsibilities: Find the volumes of cones with different slant angles.

Preparation: As a group, follow the procedures below to make several cones out of circular paper cut-outs and compare their dimensions and volumes.

Procedures:

1. To form a cone, cut out a sector from the circle. Each group member should cut out a different-sized sector using a multiple of the angle formed when 360° is divided by the number of members in your group.

2. Cut your sector out of the circle and fold the remainder into a cone. Use tape to hold it together. Examine the cones. Display the cones from the smallest sector angle to the largest. Discuss the following questions with the group and write down your answers:

   • As the sector angle increases, what happens to the base circumference of

     the cone? What happens to the height? _____

     _____

   • As the sector angle decreases, what happens to the slant height? What

     happens to the base area? _____

     _____

3. Examine the cones. Estimate which cone has the greatest volume. Record your estimate.

   _____

4. Fill the cone with the smallest sector angle with rice, and then pour the rice into a measuring cup. Record the number of ounces in the chart. Repeat this process for each cone. Which cone has the greatest volume?

   _____

5. Compare your estimate with the result obtained by measuring. Did you overestimate, underestimate, or guess correctly?

   _____

6. Compare your results with those of other groups. What size sector angle is needed to maximize the volume?

   _____

Cooperative-Learning Activity
12.6 Surface Area of Cones and Pyramids

Group Members: 2–3

Materials: different colored construction paper, scissors, ruler, paper, and tape

Responsibilities: Use forms in the shape of polygons to build as many different pyramids as possible.

Preparation: A pyramid is a solid figure whose base can be any polygon and whose faces are triangles. As a group, follow the procedures below.

Procedures: **1.** Make a pattern of the polygonal shapes by tracing or photocopying them, and then cut them out with scissors. Use different colors of construction paper for each shape, and make additional copies as indicated below each figure. Be sure to cut the shapes precisely.

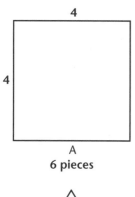

A
6 pieces

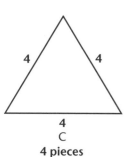

4
C
4 pieces

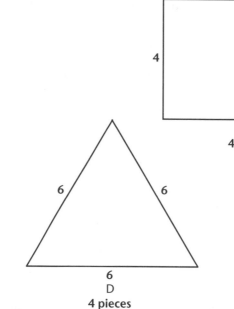

6
D
4 pieces

6

4

B
4 pieces

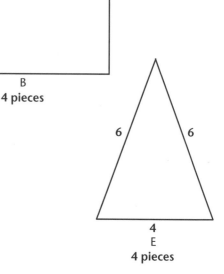

4
E
4 pieces

2. Build a pyramid from equilateral triangle D and isosceles triangle E. Name the pyramid.

_____

3. Use shapes B, C, and E to build a pyramid. Then describe your pyramid.

_____

4. Build as many different pyramids as possible from selected pieces. Draw the nets for each pyramid on a separate piece of paper. How many different pyramids are possible?

_____

Cooperative-Learning Activity
12.7 Circles and Spheres

Group Members: 3

Materials: compass, tape, and colored pencils

Roles: **Drawer** draws the circles

Folder folds and tapes the shapes

Recorder records the groups input

Preparation: Each group will construct a great circle and relate its area to the surface area of a sphere.

Procedures:
1. The members decide on their roles. Using a compass, the Drawer draws two circles, each with a radius of 3, and then cuts out each circle.

2. The Folder folds one circle in half to form two half circles and then folds it in half again to form four equal sectors. Then the Folder folds once again to form eight equal sectors, opens the folded circle, and numbers its central angles with the numbers 1 through 8. The Folder then cuts out each sector and tapes the sectors together to represent the arrangement shown. The Folder then folds the other circle in half, opens it to form a 90° angle, and tapes the rearranged circle into the half opened circle as shown.

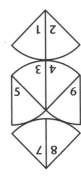

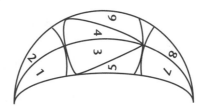

With the group's assistance, the Recorder writes answers for the following questions:

3. How much area of the sphere does your rearranged circle cover? _____

4. How many more rearranged circles would you need to build a sphere? _____

5. How is the area of a great circle, such as the one that was used to form the rearranged circle, related to the surface area of a sphere?

_____

6. Describe how you can use the area of a great circle to find the surface area of a sphere.

_____

_____

_____

ANSWERS

Cooperative-Learning Activity—Chapter 1

Lesson 1.1

Answers will vary.

Lesson 1.2

Answers will vary based on the students' heart rates.

Lesson 1.3

3. $c = 0.323f + 7.00$; $16.69

4. for 5–45 ccf, $c = 0.323f + 7.00$
for 50 ccf, $c = 0.279f + 9.00$

5. Answers will vary.

Lesson 1.4

Polygon	Number of sides
triangle	3
quadrilateral	4
pentagon	5
hexagon	6
heptagon	7
octagon	8
nonagon	9
decagon	10

Process	Number of diagonals
$\frac{3(3-3)}{2} = 0$	0
$\frac{4(4-3)}{2} = 2$	2
$\frac{5(5-3)}{2} = 5$	5
$\frac{6(6-3)}{2} = 9$	9
$\frac{7(7-3)}{2} = 14$	14
$\frac{8(8-3)}{2} = 20$	20
$\frac{9(9-3)}{2} = 27$	27
$\frac{10(10-3)}{2} = 35$	35

5. $\frac{n(n-3)}{2}$

Lesson 1.5

Answers will vary.

Lesson 1.6

5a. 64 beans **b.** Use exponents.

c. $4^{10} = 1{,}048{,}576$ **d.** after 8 days

e. Answers will vary.

Lesson 1.7

Answers will vary.

Lesson 1.8

Answers will vary.

Cooperative-Learning Activity—Chapter 2

Lesson 2.1

5, 16, 14.2

32, 8.9, 4.2

−16, −5, −32

ANSWERS

0, 16, 8.9

−340, 0, 5

Lesson 2.2

$930.00

$920.00

$870.00

$1220.00

$1108.00

$1308.00

$1278.00

$1110.00

$1051.00

$961.00

$726.00

$381.00

−$319.00

$181.00

$31.00

−$19.00

−$39.00

1. The negative ending balance indicates that the account is overdrawn by $39.00.

Lesson 2.3

Answers will vary.

Lesson 2.4

Answers will vary.

Lesson 2.5

1. a loss of 2190 Calories

2. about 10 candy bars

3. a gain of about 790 Calories

4. about 72 minutes

5. about 264 minutes

6. about 53 minutes

Lesson 2.6

A. $y = x − 5$; when $x = 4$, $y = −1$, and when $x = 5$, $y = 0$
Check students' graphs.

B. $y = 3x + 10$; when $x = 4$, $y = 22$, and when $x = 5$, $y = 25$
Check students' graphs.

C. $y = −2x + 2$; when $x = 2$, $y = −2$, and when $x = 3$, $y = −4$
Check students' graphs to make sure that they are linear and that there is a negative slope.

D. $y = −4x$; when $x = 2$, $y = −8$, and when $x = 3$, $y = −12$
Check students' graphs to make sure that they are linear and that there is a negative slope.

Lesson 2.7

1. Euclid   **2.** Pythagoras   **3.** Descartes

4. Euler   **5.** Check students' work.

Cooperative-Learning Activity—Chapter 3

Lesson 3.1

Answers will vary depending on how the cards are played out. The teacher should supervise the activity to ensure that students are playing the game correctly.

Lesson 3.2

Answers will vary depending on the rolls of the number cubes. Using the least common denominator, $\frac{11}{12}$, $\frac{5}{16}$, and $\frac{7}{24}$ can be renamed as $\frac{44}{48}$, $\frac{15}{48}$, and $\frac{14}{48}$.

Lesson 3.3

Answers will vary depending on the game boards constructed by the students. The teacher should check the game boards to ensure that each fraction has an equivalent decimal. The fractions with denominators of 3, 6, and 9 are repeating decimals.

Lesson 3.4

Answers will vary depending on the stocks chosen by the students. The teacher should help students find the necessary information in the newspaper.

Lesson 3.5

Answers will vary depending on the recipes chosen. Check students' calculations.

Lesson 3.6

Answers will vary depending on the room chosen and its size. The teacher should check the students' measurements and make sure that the students are rounding correctly.

Lesson 3.7

Answers will vary depending on the data collected. The denominator for the pooled data ratio is the total number of students in the class. The numerator for the pooled data ratio is the sum of each group's numerator.

Lesson 3.8

Answers will vary, but the experimental probability should be close to $\frac{3}{7}$.

Lesson 3.9

The theoretical probability of rolling a sum greater than 7 on any number of rolls is $\frac{3}{7}$.

The experimental probability of rolling a sum greater than 7 on 1000 rolls of the number cubes is the ratio of $\frac{\text{number of successful events}}{1000}$.

This ratio is likely to be close to the theoretical probability, $\frac{3}{7}$. The results of the activity for Lesson 3.8 may or may not be close to $\frac{3}{7}$.

Students may mention that as more trials are completed, the experimental probability gets closer to the theoretical probability.

Cooperative-Learning Activity—Chapter 4

Lesson 4.1

1. To find right angles, fold the paper in half lengthwise and again widthwise.

2. To find parallel lines, fold the paper in half once and then again; the folds will be parallel.

3. To find a 45° angle, fold the right angle in half by folding the sides of the right angle on top of one another.

4. a 135° angle is a 90° angle plus a 45° angle

5. a 225° angle is a 135° angle plus another 90° angle

Lesson 4.2

The answers will depend on the angles cut out by the instructor.

Lesson 4.3

The students' conjectures should include the following:

Alternate interior angles are congruent.
Alternate exterior angles are congruent.
Consecutive interior angles are supplementary.
Corresponding angles are congruent.

ANSWERS

Lesson 4.4

A regular tessellation can be constructed only from regular triangles, squares, or regular hexagons.

Lesson 4.5

Answers will vary depending on the objects chosen by the teacher. A square with the same perimeter as each rectangle will have the greatest possible area. Each side of the square is one-fourth of the perimeter of the rectangle.

Lesson 4.6

Answers will vary depending on the figures constructed by the students.

Lesson 4.7

The figure is a spiral whose outside edges are the 1-inch legs of the triangles. The next three hypotenuses would have lengths of $\sqrt{14}$, $\sqrt{15}$, and $\sqrt{16}$, or 4. The figure is called a radical spiral because all of the hypotenuses of the triangles are radicals and they look like a spiral as they wind around the center.

Lesson 4.8

Answers will vary depending on the figure drawn by the student and the scale factor chosen. Check students' drawings. The hypotenuses are related by the same scale factor as the legs. The corresponding angles of the triangles are congruent. The shape of the original triangle is preserved because the scale factor preserves the angle measures of the triangle.

Cooperative-Learning Activity—Chapter 5

Lesson 5.1

1. a. Nevada **b.** Arizona **4. a.** $4a - 2$

b. $8x - y$ **c.** $9x + 3y + 3a + 2$ **d.** $3x$

e. $4a - 6b$ **f.** $12a + 4b$

g. $-4x + 5y + 2$ **h.** $-8x + 1$

Lesson 5.2

2. a. $-x - 3$

  b. $-2y + 2$

  c. $x + y$

  d. $-2x - y$

  e. $-2x - 3y$

  f. $2x - 3y$

  g. -10

  h. -4

4. true; false

Lesson 5.3

Answers will vary. Solutions depend on the squares chosen.

Lesson 5.4

Answers will vary. Solutions depend on the rolls of the number cube.

Lesson 5.5

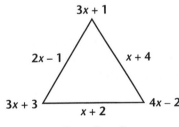

Sum: $8x + 3$

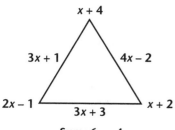

Sum: 6x + 4

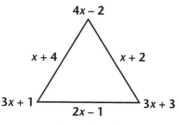

Sum: 8x + 3

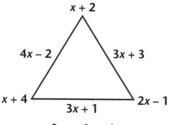

Sum: 6x + 4

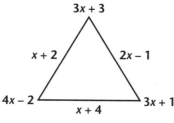

Sum: 8x + 3

Lesson 5.6

Answers will vary. Each inequality of the form $a < x$ is equivalent to $x > a$. Each inequality of the form $a > x$ is equivalent to $x < a$.

Lesson 5.7

Answers will vary.

Cooperative-Learning Activity—Chapter 6

Lesson 6.1

2. The number of border squares is 16.

3. $4(5 - 2) + 4; (5 \times 4) - 4$

4. One method subtracts two corner squares from each side, or a total of eight corner squares, so four of the corner squares must be added again. The other method uses two corner squares on each side, so four of the corner squares must be subtracted.

5. The number of border squares is 28; the expressions are $4(8 - 2) + 4$ and $(8 \times 4) - 4$.

6. $4(n - 2) + 4; 4n - 4$

7. $4(n - 2) + 4 = 4n - 8 + 4 = 4n - 4$

Lesson 6.2

$a = -3, x = 1, b = -2, n = -5, c = 4$

Lesson 6.3

1. Possible answer:

Number	124
Multiply by 7	868
Multiply by 11	9548
Multiply by 13	124,124

2. Answers will vary.

3. 1001; 527,527; 111,111; 989,989

$527 \times 1001 = 527 \times (1000 + 1) =$
$527 \times 1000 + 527 \times 1 =$
$527,000 + 527 = 527,527$

4. Write the number. Repeat the digits in the same order to make a six-digit number.

5. Switching the order in which you multiply by the factors would not change the final result because 1001 would still be the other factor of the product.

ANSWERS

Lesson 6.4

2. Let x represent the number of boys who can tug as hard as 1 girl.

$$\frac{4}{5} = \frac{x}{1}; x = \frac{4}{5}$$

3. $1G = \frac{4}{5}B$

4. $1D = 2G + 1B$

$$= 2\left(\frac{4}{5}\right)B + 1B$$

$$= \frac{8}{5}B + 1B$$

$$= 2\frac{3}{5}B$$

$2\frac{3}{5}$ boys can pull as hard as 1 dog.

5. The dog and three girls will win.

Lesson 6.5

Answers will vary.

Lesson 6.6

1–3.

	Occurrences	Percent
A	11	8.8
B	1	0.8
C	3	2.4
D	4	3.2
E	20	16
F	3	2.4
G	3	2.4
H	7	5.6
I	8	6.4
	Total: 60	

	Occurrences	Percent
J	0	0
K	0	0
L	8	6.4
M	0	0
N	9	7.2
O	7	5.6
P	2	1.6
Q	0	0
R	6	4.8
	Total: 32	

	Occurrences	Percent
S	15	12
T	11	8.8
U	2	1.6
V	3	2.4
W	2	1.6
X	0	0
Y	0	0
Z	0	0
	Total: 33	

4. E, S, and A or T

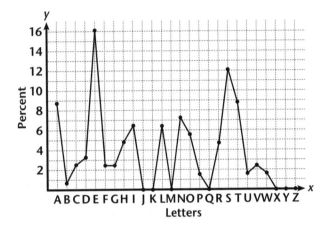

Cooperative-Learning Activity—Chapter 7

Lesson 7.1

2. The result will display the day, month, and year of birth of each member.

3.

Step 1	m
Step 2	$20m$
Step 3	$20m + 3$
Step 4	$100m + 15$
Step 5	$100m + 15 + d$
Step 6	$2000m + 300 + 20d$
Step 7	$2000m + 303 + 20d$
Step 8	$10{,}000m + 1515 + 100d$
Step 9	$10{,}000m + 1515 + 100d + y$
Step 10	$10{,}000m + 100d + y$

ANSWERS

4. In the final number, the digits with the place values 10^5 and 10^4 represent the month, the digits with the place values 10^3 and 10^2 represent the day, and the digits with the place values 10^1 and 10^0 represent the year of birth; to cancel the 1515 that was added in Steps 3–7; $b = 10,000m + 100d + y$

Lesson 7.2

1–3. The final result displays the numbers rolled on the number cube in the order they were rolled.

4. Step 1 $2a$
   Step 2 $2a + 5$
   Step 3 $10a + 25$
   Step 4 $10a + b + 25$
   Step 5 $100a + 10b + 250$
   Step 6 $100a + 10b + c + 250$
   Step 7 $100a + 10b + c$

5. The first roll is displayed by the hundreds digit, the second by the tens digit, and the final roll by the ones digit; to cancel the 250 that was added in Steps 2–5; $n = 100a + 10b + c$

Lesson 7.3

2. **a.** $p = b$ **b.** $p = c + s$ **c.** $3s = 2b$

3. Since $p = b$, $2p = 2b$; but $3s = 2b$ and $2p = 2b$, so $3s = 2p$

   $p = c + s$
   $3p = 3c + 3s$
   $3p = 3c + 2p$
   $p = 3c$
   Thus, 3 cups will balance a pitcher.

Lesson 7.4

1.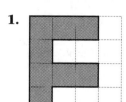

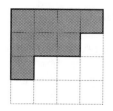 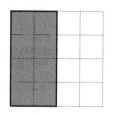

2. The greatest perimeter is 18.

3.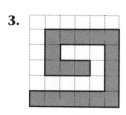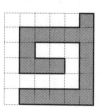

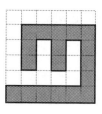 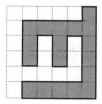

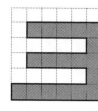 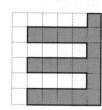

4. 18 square units; all cuts have a perimeter of 38.

Lesson 7.5

2. A: Each group will form a triangle.
   B: The measurements 3 cm, 4 cm, and 8 cm will not form a triangle.
   C: Each group will form a triangle.

3. 3 cm, 5 cm, 8 cm; 2 cm, 4 cm, 5 cm; 6 cm, 6 cm, 6 cm; 6 cm, 7 cm, 6 cm; 4 cm, 5 cm, 7 cm; 4 cm, 4 cm, 7 cm; 4 cm, 5 cm, 8 cm

4. To form a triangle, the sum of the lengths of two sides must be greater than the length of the third side.

Lesson 7.6

5. 1, 7; −2; 2, 6

ANSWERS

6. Move the Distance From scale so that the 0 aligns with the 1 on the Number Line scale. Find the two 4s on the Distance From scale and read the numbers on the Number Line scale that are directly below and between the two 4s on the Distance From scale; $x > -3$ and $x < 5$

7. Absolute-value equations have a finite number of solutions. Absolute-value inequalities may have an infinite number of solutions.

Cooperative-Learning Activity—Chapter 8

Lesson 8.1

1.

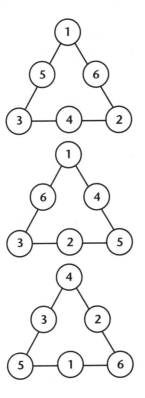

2, 3.

Sum of the sides (s)	9	10	11	12
Sum of the vertices (v)	6	9	12	15
Sum of the numbers in the triangle	21	21	21	21

4. A triangle has three sides.
Two times; three times the sum along each side minus the sum of the vertices equals the sum of the numbers used in each triangle, or $2s - v = 21$.

5. There is a linear relationship between the sum of the sides and the sum of the vertices because as the sum of the sides increases by 1, the sum of the vertices increases by 3.

Lesson 8.2

1. The slope of the segment connecting (1, 4) and (2, 8) is 4.
The slope of the segment connecting (2, 8) and (3, 2) is −6.
The slope of the segment connecting (3, 2) and (4, 6) is 4.
The slope of the segment connecting (4, 6) and (5, 0) is −6.
The slope of the segment connecting (5, 0) and (6, 4) is 4.
The slope of the segment connecting (6, 4) and (7, 8) is 4.

2. The positive slopes are 4; the negative slopes are −6; the slope is not constant.

3. The graph repeats in a cycle of 5 units.

4.

Whole number	Slopes of segments	Length of cycle
3	−7, 3	10
5	−5, 5	2
6	−4, 4, 6	5
7	−3, 3, 7	10

5. The slope of the next segment on the 3's graph is 3.
The slope of the next segment on the 5's graph is 5.
The slope of the next segment on the 6's graph is 6.
The slope of the next segment on the 7's graph is −3.

ANSWERS

Lesson 8.3

1.

Alissa	
time in seconds (t)	distance in meters (d)
1	6
2	12
3	18
4	24
5	30

John and Alissa		
John's time in seconds (t)	John and Alissa's combined distance in meters (d)	John and Alissa's combined time in seconds
5	37.5	10
8	42	13
10	45	15
15	52.5	20
20	60	25

Andy	
time in seconds (t)	distance in meters (d)
0	0
5	15
10	30
15	45
20	60

2.

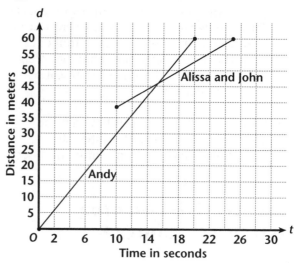

3. John begins running after Alissa has completed 30 meters; 30 m; 60 m; Andy wins the race because he runs 60 m in 20 s while John and Alissa run 60 m together in 25 s.

4. For $d = 60$, Andy's graph is to the left of John and Alissa's graph.

Lesson 8.4

1. Answers will vary.

2. The other winning arrangements are (1, 1), (2, 1), (3, 1); (1, 2), (2, 2), (3, 2); (1, 3), (2, 3), (3, 3); (1, 1), (1, 2), (1, 3); (2, 1), (2, 2), (2, 3); (3, 1), (3, 2), (3, 3); and (1, 1), (2, 2), (3, 3).

3. The equations are $x = 1$, $x = 2$, $x = 3$, $y = 1$, $y = 2$, $y = 3$, $y = x$, and $y = -x + 3$.

4. The slopes of the diagonals are 1 and -1. The slope of the horizontal lines is 0 because there is no change in the y-values of the points.

Lesson 8.5

Answers will vary.

Lesson 8.6

1. Check students' geoboards.

ANSWERS

2. 0; the run from $(-2, -1)$ to $(2, -1)$ is 4; the rise is 0; $\frac{0}{4} = 0$

3. 0; the rise from $(-2, 2)$ to $(2, 2)$ is 0; the run is 4

4. The rise of a horizontal line is 0.

5. The rise from $(-2, 1)$ to $(-2, -2)$ is 3; 0; the slope of the line is undefined

6. Answers will vary.

7. Division by 0 is undefined.

8. 2; 0; $\frac{1}{2}$; the y-intercept; $y = 2x$ and $y = \frac{1}{2}x$

Lesson 8.7

1. The segments are the same length.

2. q; The slope of each line segment is 2.

3. The segments are equal in length; the slopes are 4 and -4; because the top horizontal line is closer to the two slanted lines, it appears longer than the bottom horizontal line.

Lesson 8.8

Variables will vary. For example, let x represent the number of hazelnut-filled chocolates.

1, 3.

Type of filling	Variable expression	Number
hazelnut	x	4
raspberry	$2x$	8
almond	$x - 2$	2
coconut	$2x - 3$	5

2. $x + 2x + x - 2 + 2x - 3 = 19$; $x = 4$

5. Answers will vary.

6. Let n represent the number of selections; $n \leq 5$.

Cooperative-Learning Activity—Chapter 9

Lesson 9.1

1.

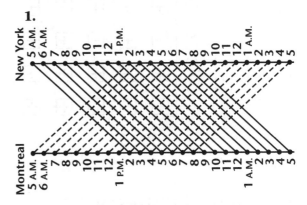

2. The slope of the lines in each set are equal. Since each van takes 9 hours to travel between the two cities, the rate of speed is the same. Therefore, the slope is the same.

3. The points of intersection represent the number of vans that each van passes on its way between New York and Montreal.

4. The van will pass 9 vans along the way. When the van arrives in Montreal, it will pass a van that is leaving. Therefore, the van will pass a total of 10 vans.

5. Answers will vary.

Lesson 9.2

1.

Owes	Jake	Oko	Luc	Emily
Jake	0	0	0	1
Oko	0	0	0	0
Luc	0	0	0	0
Emily	0	0	0	0

2. The table shows that once Jake pays Emily back the dollar, all the debts have been paid.

3–4. $j = -o, o = -l, l = -j, j = -e$

5. $j = 0, o = 0, l = 0, e = 0$

6. Answers will vary.

Lesson 9.3

2. $\begin{cases} x + y = 20 \\ 3x - y = 0 \end{cases}$

3.

4. $4x = 20$

5. $x = 5$

6. $x = 5$ and $y = 15$; there are 5 correct answers.

Lesson 9.4

2–3. For each system, the solution is $x = -1$ and $y = 2$.

4. In each equation, the difference between the x- and y-coordinates is equal to the difference between the y-coordinate and the constant.

5. Answers will vary.

Lesson 9.5

1. A: $y = x$; $\begin{cases} x \le 4 \\ y \ge 0 \\ y \le x \end{cases}$

2. B: $\begin{cases} y \le 3 \text{ and } y \ge 0 \\ x \ge 0 \text{ and } x \le 9 \end{cases}$

C: $y = x + 2$; $\begin{cases} x \ge 0 \\ y \le x + 2 \\ x \le 2 \\ y \ge 0 \end{cases}$

D: $\begin{cases} x \le 4 \\ x \ge 0 \\ y \le 8 \\ y \ge 0 \end{cases}$

3. Answers will vary.

Cooperative-Learning Activity—Chapter 10

Lesson 10.1

2. Answers will vary. This experiment represents exponential decay. The curve should resemble an exponential function.

Lesson 10.2

1. Check students' sketches.

2–3. Answers will vary.

Lesson 10.3

2.a. The reciprocal function is
$f(x) = \dfrac{1}{x} + 2$, where $x \ge 1$.

b. The reciprocal function is $f(x) = \dfrac{1}{x^2}$, where $x \ge 1$.

c. The reciprocal function is $f(x) = \dfrac{1}{3}$.

3. It appears that as the x-values increase, the y-values decrease, and as the x-values decrease, the y-values increase. However, the horizontal line $y = 3$ forms a horizontal line of $y = \dfrac{1}{3}$, which does not fit this description.

Lesson 10.4

2. The graph represents a step function.

Lesson 10.5

1–3. Answers will vary.

4. If n is the number of counters of one color, the number of moves is represented by the expression $(n + 1)^2 - 1$.

Lesson 10.6

1. Rotations: 1-3-5-2-4 3-5-2-4-1
5-2-4-1-3 2-4-1-3-5 4-1-3-5-2

2. Reflections: 4-2-5-3-1   1-4-2-5-3
3-1-4-2-5   5-3-1-4-2   2-5-3-1-4

There are 10 solutions.

3a. 4 solutions

b. 5 solutions

Cooperative-Learning Activity—Chapter 11

Lesson 11.1

Answers will vary depending on the contents of each box of animal crackers.

Lesson 11.2

Answers will vary.

Lesson 11.3

Answers will vary.

Lesson 11.4

1. Starting on the color cranberry, when you have spelled out the letters in a color, you should end up on that color.

2. The result is the same for all colors.

Lesson 11.5

5a. The better the estimate, the more closely the points fit a straight line.

5b. Estimation skills should improve as the activity progresses.

Lesson 11.6

1–5. Answers will vary.

Cooperative-Learning Activity—Chapter 12

Lesson 12.1

1–4. Answers will vary. The more accurate the estimated area, the more closely the data points fit the parabolic curve.

Lesson 12.2

3. The surface is a rectangle.

4. Answers will vary. The container must be tilted on a corner.

5. Answers will vary. An octagon and a right triangle are not possible.

Lesson 12.3

1. Answers will vary.

2. The smallest number of colors is 3.

3. Answers will vary.

4–6. Answers will vary.

6.

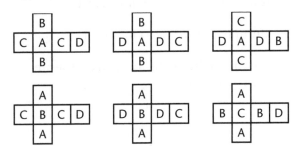

7. Six solutions with exactly 4 colors are possible.

Lesson 12.4

3–7. Answers will vary. As estimation skills increase, the data points approach a correlation of 1.

ANSWERS

Lesson 12.5

2. As the sector angle increases, the circumference decreases, and the height increases. As the sector angle decreases, the slant height increases.

3–5. Answers will vary.

6. Answers will vary. The group should notice that a sector of 0° has no volume, then the volume increases. However, when the sector angle is 360°, the volume again nears 0.

Lesson 12.6

2. The pyramid is a triangular pyramid.

3. The pyramid is a rectangular pyramid.

4. Rectangular pyramids: ACCCC, AEEEE, BDDEE, BCEEE
Triangular pyramids: CCCC, DDDD, CEEE, DDEE, EEEE

Lesson 12.7

3. The great circle represents one-fourth of the area of the sphere.

4. You would need three more great circles.

5. Multiply the area of a circle by 4.

6. You can multiply the formula for the area of a circle by 4.